YOU ARE HERE

YOU ARE HERE

A Portable History of the Universe

Christopher Potter

HARPER

An Imprint of HarperCollins*Publishers*
www.harpercollins.com

HarperCollins books may be purchased for educational, business, or sales
promotional use. For information, please write: Special Markets Department,
HarperCollins Publishers, 10 East 53rd Street, New York, NY 10022.

First published in Great Britain in 2009 by Hutchinson, an imprint of
Random House Group Limited.

FIRST U.S. EDITION

Library of Congress Cataloging-in-Publication Data is available
upon request.

ISBN 978-0-06-113786-0

09 10 11 12 13   OFF/RRD   10 9 8 7 6 5 4 3 2

For my mother

Contents

Do I dare
Disturb the universe?

    T. S. Eliot, 'The Love Song of J. Alfred Prufrock'

Orientation

The eternal silence of these infinite spaces alarms me.
                                                    Blaise Pascal

You are here, it says on the map in the park, the train station and the shopping centre, an arrow, usually red, pointing to some reassuringly definite location. But where is here, exactly? Children know, or think they know. On the flyleaf of a first book, I wrote out, as we all did after our own fashion, my full cosmic address – Christopher Potter, 225 Rushgreen Road, Lymn, Cheshire, England, The United Kingdom, The World, The Solar System, The Galaxy – my childish handwriting getting larger and larger, as if each part of the address I knew to be bigger and more important than the preceding part, until, with a final flourish, that acme of destinations is reached: the universe itself, the place that must locate everything there is.

As children we soon become aware that the universe must be a strange place. I used to keep myself awake at night trying to imagine what lay beyond the edge of the universe. If the universe contains everything there is, then what is it contained in? We now know, scientists tell us, that the visible universe is a region of radiation that evolved and is not contained in anything. But such a description raises too many questions that are more disturbing than the question we had hoped to have had answered in the first place, and so we quickly put the universe back in its box and think about something else instead.

We do not like to think about the universe because we fear the immensity that is everything. The universe reduces us to a nub, making it difficult to escape the idea that size matters. After all, who can deny the universe when there is so much of it? 'Spiritual aspirations threaten to be swallowed up by this senseless bulk into a sort of nightmare of meaninglessness,' wrote the Anglo-German scholar Edward Conze (1904–1979). 'The enormous quantity of matter that we perceive around us, compared with the trembling little flicker of spiritual insight that we perceive within us, seems to tell strongly in favour of a materialistic outlook on life.' We know that we must lose if we are to contest the universe.

Just as terrifying is the idea of nothing at all. A little while ago each of us was nothing, and then was something. No wonder children have nightmares. The something of our existence ought to make the nothingness that preceded life an impossibility, since we also know, as King Lear observes, that 'nothing can come of nothing'. And yet every day in the annihilation and miraculous resurrection of the ego that is going to sleep and waking up, we are reminded of that very nothingness from which each of us emerges.

If there is something – which there appears to be – then where did that something come from? Such thoughts coincide with the first inklings we have of our own mortality. Death and nothingness go hand in hand: twin terrors to put alongside our terror of the infinite; terrors we spend the rest of our lives suppressing into the shape of our adult selves.

Humans are caught in a bind. On the one hand we know that there is something because we are each sure of our own existence; but we also know there is nothing because we fear that that is where we came from and where we are headed. We know intellectually that the nothingness of death is inescapable but do not actually believe it. 'We are all immortal,' the American novelist John Updike reminds us, 'for as long as we live.'

'What happens when I die?' a child soon asks, a question that as adults we also put to one side. Not even a material girl in a material world would be satisfied with an answer that was restricted to descriptions of physical decay, and yet even a material answer to such a question, and indeed to all questions, will end up at the same place. What is the material of the world and where does it come from? To think about the universe is to ask again the childhood questions we no longer ask: What is everything? And what is nothing?

Seemingly all children start out as budding scientists unafraid to follow a trail of questioning to exhaustion, even if the exhaustion is usually that of their parents. Curiosity drives children to ask why? And why? And why? hoping to arrive at some final destination, like the universe at the end of our cosmic address, a final answer beyond which there are no more whys.

'Why is there something rather than nothing?' asked the German philosopher Gottfried Leibniz (1646–1716), the question that any description of the universe must ultimately be able to address. Science attempts to answer 'why' questions with 'how' answers, invoking the dynamic of stuff in the world. But 'how' answers also converge on that same ultimate question: instead of asking *why* is there something rather than nothing?', scientists ask *how* did something come out of nothing?' In order to account for the everything-ness of the universe we must also account for the nothing-ness from which it seems to have appeared. But what could such material as the world is made of look like when it is nothing, and what possible actions could have transformed nothing into something, and something into the everything we call the universe?

For hundreds of years, and for as long as the word has meant anything, science has shown itself to be an evolving process of investigation into whatever it is that is Out There, a place of things that are in motion, and what we mean by the universe. So who better, we might think, than scientists to answer the question: Where – between the void and everything – are we?

Their replies are not always encouraging:

- 'Man knows at last, that he is alone in the unfeeling immensity of the universe, out of which he has emerged only by chance,' the French biologist Jacques Monod (1910–1976) once wrote, with what sounds like glee that we should have finally found this out.
- 'Science has revealed much about the world and our position in it. And generally, the findings have been humbling.' writes Nick Bostrom, Director of the Future of Humanity Institute at the University of Oxford. 'The earth is not the centre of the universe. Our species descended from brutes. We are made of the same stuff as mud. We are moved by neurophysiological signals and subject to a variety of biological, psychological and sociological influences over which we have limited control and little understanding.'
- 'Our true position,' says the American physicist Armand Delsemme, '[is one] of isolation, in an immense and mysterious universe.'

Isolated in pointlessness: no wonder we non-scientists would rather stay indoors and watch television, or read *Middlemarch*, or do whatever it is that we do indoors. If this is the universe as science describes it, then we surely want none of it. Such a description only reignites those nauseous existential fears we have suppressed since childhood.

Or are these my fears and not yours? I have friends who claim that they never think about the universe at all. And yet I can't help but feel that such rejection – of the universe of all things! – is evidence of deep repression rather than lack of interest. Who, after all, wants to be told that they are insignificant specks in a vast, purposeless and uncaring universe? And if we do take note, it's hard not to blame science for finding it out. These stark scientific pronouncements seem impossible to deny. Easier, then, not to think about science either, for fear of being told something irrefutable that we would much rather not know: that we do not have free will; that the mind is merely a quality of the brain; that gods do not exist; that the only reality is material reality; that any knowledge that isn't scientific knowledge is not just worthless, it isn't knowledge at all.

Sometimes it seems that what science is telling us is that the universe has little in common with the subjective experiences that define us as human beings. We seem to be in opposition to a universe at best uninterested in the qualities that make us human, which makes some of us think – a thought we'd presumably rather not have – that to be human is to be intrinsically separate from the source of our own creation.

To be at peace with the universe is not easy. The English mathematician Frank Ramsey (1903–1930) found a way of accommodating the universe by accommodating the idea of size itself: 'Where I seem to differ from some of my friends is in attaching little importance to physical size. I don't feel the least humble before the vastness of the heavens. The stars may be large, but they cannot think or love; and these are qualities which impress me far more than size does . . . My picture of the world is drawn in perspective . . . The foreground is occupied by human beings, and the stars are all as small as threepenny bits.' The contemporary astronomer Alan Dressler has a similar strategy: 'If we could learn to look at the universe with eyes that are blind to power and size, but keen for subtlety and complexity, then our world would outshine a galaxy of stars.'

Drawing the universe in human scale might remind us of the world as seen in paintings before the discovery of formal perspective, where a different hierarchy of size is imposed. In pre-Renaissance paintings, the hierarchy is based on relative spiritual importance, so that the Virgin Mary, say, looms large over the saints, who in turn dominate the kneeling donor who has commissioned the painting in the first place. For Ramsey it is humanity that is the measure of the world, not a spiritual nor a literal yardstick. But this doesn't help us much if, putting aside all fears and existential vertigo, we cannot escape the idea that science might be all that there is, that the whole universe can be measured and brought to account. We might all too easily convince ourselves that science reduces our lives to files and card indexes, as in some totalitarian regime that believes its citizens are best subdued when they are reduced to statistics. Rigid, authoritarian, patriarchal, analytical, without emotional content: these are some of the qualities we could be tempted to ascribe to science and scientists.

But there is another side. Half a century ago, the English astronomer and physicist Fred Hoyle (1915–2001) noted, admittedly with a hint of exasperation, the curious fact that 'while most scientists claim to eschew religion, it actually dominates their thoughts more than it does the clergy'. Certainly most of the prominent scientists of the past were believers. A recent poll shows that perhaps 50 per cent of scientists today believe in some form of a personal God, while another poll tells us that only 30 of the hundred physicists who were asked believe that parallel universes actually exist. 'I would like to know how God created the world,' Einstein[1] once said. 'I am not interested in this or that phenomenon, in the spectrum of this or that element. I would like to know His thoughts. The rest is detail.'

Even hard-line materialists like the English theoretical physicist Stephen Hawking (b.1942) and the American physicist Steven Weinberg (b.1933) sprinkle their writings with talk of the possible nature of the God in which they do not believe. Hawking tells us that we may actually be close to knowing the mind of God, while Weinberg, even-handedly, tells us that 'science does not make it impossible to believe in God. It just makes it possible to not believe in God.'

Science is atheistic only in so far as it means to explain nature without recourse to the supernatural. In science, nature can be mysterious but it is not permitted to be mystical. *Scientists*, however, need not be atheistic, nor must agnosticism necessarily rule out spirituality. If science ever explains everything, only then do the gods die. But can science ever explain every-

---

[1] There has been much speculation about Einstein's religious views. It appears that he did not believe in a personal God, but his views are perhaps best understood by reference to what he actually said. The word God is dotted about his writings.

thing? Hawking has proclaimed that 'we may now be near the end of the search for the ultimate laws of nature', but it is far from clear that this is the case. At the end of the nineteenth century a similar declaration was made by the American physicist Albert Michelson (1852–1931): 'It seems probable that most of the grand underlying principles have been firmly established and that further advances are to be sought chiefly in the rigorous application of these principles to all phenomena which comes under our notice.' He could not have been more wrong. One of the most fertile periods in the history of science was just about to begin. The universe's finest joke may be to reveal itself, as science systematically uncovers some of its secrets, as ever more mysterious.

In any case, since science has persuaded us to be agnostic about almost everything, perhaps now, in the ultimate act of modern ennui and irony, we might be inclined to be agnostic about science too. 'Your cry of triumph at some new discovery will be echoed by a universal cry of horror,' the German playwright Bertolt Brecht (1898–1956) has Galileo say in his play *The Life of Galileo*. What is the cost of knowledge, we ask more and more insistently as science both creates and brings to the edge of destruction the world we live in? Sometimes the very certainty of the uncertainty that science has uncovered looks like dogmatism. Why do I feel sure that the uncertainty some scientists urge us to embrace is not what the poet Keats had in mind when he wrote of the 'Man of Achievement . . . capable of being in uncertainties, Mysteries, doubts, without any irritable striving after fact & reason', a quality he named Negative Capability? For the same reason, I suspect, that I am disturbed by the wild-eyed optimism of those scientists who urge us to look to further scientific progress to put a damaged world back together again.[2] How much unrestrained scientific optimism in unrestrained scientific progress can we bear?

---

[2] It has been suggested that we might reduce global warming by injecting sulphur dioxide into the upper atmosphere, or by pumping cold water from the bottom of the ocean to the surface.

The scientific method, like capitalism, is always in search of new territory to exploit. Capitalism, Marx predicted, would come to an end when there were no more markets left. In our own age, the emergence of some of the largest markets in the history of civilisation makes such an end seem a long way off. And science outstrips even capitalism. We have begun to realise that there may not be so much time left for the earth, at least not as a place willing to host us. Not to worry, say the champions of scientific materialism, trust us, we are certain (well, pretty certain) that when we have conquered space we will find that there are many other places, somewhere out there, that we might claim as home. And if there aren't, we'll just build you a new one from scratch.

But for all the confident talk of leaving home and finding other places to live, such far-flung travel is highly speculative, hardly proper science even, given the limits set by our current understanding of the laws of nature. Perhaps the more we know about how the universe is constructed, the more reasons we will discover why we are bound to this place as our home. Putting aside all the hopes of science fiction, and science theory so speculative it might as well be fiction, it seems more realistic to suppose that we are unlikely ever to travel beyond the solar system, perhaps not even so far. Mankind hasn't walked on the moon for over a generation, and we are just beginning to realise that even such short astronomical hops can cause considerable psychological trauma. It is far from clear what we would have to become – some sort of man-made post-human form perhaps – in order to live somewhere else. It may be that we are peculiarly suited to the earth, and such knowledge might force us to take better care of it. In 2006, Stephen Hawking wrote that the best hope for the future survival of mankind would be to abandon the earth and look for a new home. In the meantime, it might be a good idea to have a Plan B.

I want to know what this universe is that attracts and repels me, and which is described by a methodology that also attracts and repels. I am attracted to science for its power, beauty and mystery, and its call to live in uncertainties; I am repelled by its power, nihilism and smug material certainty. Perhaps these polar extremes might be reconciled if I can begin to understand what it is that scientists are doing when they do what they do.

At school, the relationship between science and nature (the universe as it appears on our doorstep) never came into it. I'm not even sure I ever made a connection between what went on in the laboratory and what goes on in the natural world as it manifests itself around us. In physics, the world was simulated with ball bearings and electrical hardware (where are they in the forests and the highlands?); in chemistry, we peered at reactions between chemicals that are hardly ever found out in the open; and biology, which purported to be about the living world, seemed to be more about cutting up things specially killed for the purpose. Science appeared to be about beating a reluctant world into some sort of submission. And then there was mathematics, how did that fit in? I once heard someone claim it as the queen of the sciences, but what did that mean? I gathered, somehow, that mathematics was meant to underpin science in some way, but no one in the maths department – where mathematics was thought to be too grand to have anything to do with the laboratory – was letting on.

My experience of science at school was traumatic enough to make me feel like an outsider, but not sufficiently traumatic to snuff out entirely my outsider interest in what it is that science does. It's not difficult to feel outside science: even scientists can be excused for feeling excluded. Long gone are the days when 'the laws of the universe were something a man might deal with pleasantly in a workshop set up behind the stables'.[3] Rocket-launched observatories and particle accelerators that cost billions of dollars and take years to construct have put paid to the wider democracy of science.[4] Mathematicians have always made up an exclusive club but even that club

---

[3] The German-born English novelist Sybille Bedford (1911–2006): *A Legacy* (1956).
[4] There is, in fact, still plenty of science that can be done in a shed, but the pursuit of the laws of the universe has become expensive.

is now broken up into sometimes tiny splinter groups. There are mathematical proofs that take years to check, and these are only understood by the handful of mathematicians involved in the process of verification or by those who constructed the proof in the first place. If scientists themselves are entitled to feel excluded, how much more do we, poor puzzled onlookers, peer through a glass darkly?

At school I discovered I had a modest talent for mathematics. It was Miss Church, the mathematics teacher, who educated[5] me: literally brought something forth, the opposite of the process that forces information in and is sometimes mistaken for education. So it was maths for me at university, a subject to which, I soon realised, I was never going to make an original contribution. Being OK at mathematics is like being an OK cook or an OK pianist: a yawning chasm separates the amateur and the professional. The truly gifted begin beyond the point where amateurs leave off. Good meals may result from slavish devotion to a recipe, but where are the new recipes to come from? Although there was a time when I could produce Einstein's relativistic equations, or prove Gödel's theorem from scratch, I had little idea what it was that I was doing when I reconjured these profound insights into the nature of reality. After years of education, I was no closer to understanding what it is that scientists do when they do what they do. Part of the problem is that most scientists are happy just to *do* what it is that they do without questioning what that is exactly. They are not interested in philosophical conundrums, to which their likely response is, as the American physicist Richard Feynman (1918–1988) wittily put it, 'Shut up and calculate!' Scientists are pragmatists[6]. If it works, philosophical considerations are superfluous. The American theoretical physicist Lee Smolin (b.1955) goes further. He has declared that 'in science we aim for a picture of nature as it

---

[5] Educate: from the Latin words 'e' (from) and 'ducere' (to lead).
[6] Like Gwendolen in *The Importance of Being Earnest* (1895): 'Ah! that is clearly a metaphysical speculation, and like most metaphysical speculations has very little reference at all to the actual facts of real life, as we know them.'

really is, unencumbered by any philosophical or theological pre-judice'.[7] But how can science be divorced from philosophy and theology, as if a poisoned river runs between it and other forms of enquiry? Historically, science developed out of philosophy and creation stories, and what science now knows *is* our modern creation story. In that river is exactly where I want to be.

I went back to university for what turned out to be a last gasp of formal education: a course in the history and philosophy of science, begun as a doctorate but soon abbreviated into a single year. My strongest memory of that year is of a remark made by the head of department, which I remember partly because he immediately disavowed it, and partly because I associate it with my continued feeling that I was on the outside of the world I meant to inhabit. He wondered what it would be like to teach the piano knowing that the only two physical variables are the speed and the force with which the keys are hit. Pausing for a moment, he wondered if, perhaps, there might in fact be only one variable – force, alone – given that the action of the piano is fixed. My heart leapt with interest. Here was a possible bridge across the river. 'But we stray into aesthetics,' the professor concluded, and changed the subject. And so at the end of the year I took away my Master's degree and ventured, not much wiser, out into the wider world.

Eventually, I settled down as an editor working with various writers, some of whom wrote about science and others about the vicissitudes of the human heart. And for a long time I was happy enough to have found an accommodation between two worlds.

Like many who have come to writing latish in life, it took a crisis[8] to get me here. I realised that I could either go on trying to find someone to write the book I wanted to read, or I could write it myself. Being an outsider might even be something I could turn to my advantage.

---

[7] *New Scientist*, 23 September 2006.
[8] Crisis: a turning point, a time of distress. Greek, from *Krinein*, to decide.

Is it possible for a layman to find a path through the universe science describes? I hope so. We do not feel so excluded from any other of mankind's truth-seeking enterprises. We may or may not understand some of the products of contemporary art, but at least we feel entitled to an opinion. 'I could do better myself at home' is never a response to the latest scientific theory, but perhaps we might be more inclined to venture an opinion on, say, the Large Hadron Collider, if we knew a little about what a particle accelerator is and what this one might achieve. We might even be *entitled* to an opinion given its cost, not just in dollar terms but to our current physical descriptions of reality. There are, of course, places to go to find out such information – specialist magazines and designated pages in certain newspapers – but my imagined reader feels excluded even there. She longs to take a walk across the universe but does not know from where the journey sets out, let alone where such a journey might end. She does not have the benefit even of my limited scientific background, but shares my desire to find out what science does, and is drawn, as I am, to what it is that science has to tell us of the world out there, no matter how painful such knowledge might turn out to be. Scientists have dared to venture out into the universe for centuries, armed only with a clock and a ruler. Perhaps that's why madness is a quality particularly associated with these fearless adventurers. With these magic wands in hand we can follow, not too cautiously, but cautiously enough to avoid madness and confidently enough to live up to T. S. Eliot's maxim: 'Only those who risk going too far can possibly find out how far they can go.'

26 Degrees of Separation

Man is the measure of all things: of things which are, that
they are, and of things which are not, that they are not.

Protagoras

If we are to find out where we are in the universe, we will
need to know what things are in it, and where. Scientists
measure things in metres, and so it is with a metre ruler that
we will set out. We will see what we can find, and if we are
going to be overwhelmed by the size of the universe we might
at least find out where nausea sets in.

We would make slow progress if we were to measure the
universe a metre at a time. Such cautiousness would soon turn
to boredom. We can explore further, faster, if we allow each
step to magnify tenfold – what scientists call an order of magnitude.
All objects measuring between 1 and 10 metres long fall
into one order of magnitude, which is our first step. The next
step of our walk across the universe measures those things
that are between 10 and 100 metres long, and so on. Setting
out from here where we live, we can search out the parts of
that address we sought as children, when we were scarcely
more than the height of our ruler.

1–10 metres (10^0–10^1 metres)

Between most humans there is very little variation in height.
John Keats was 1.54 metres (5 feet 3/4 inches), Admiral Lord
Nelson and Marilyn Monroe were both 1.65 metres (5 feet 5
1/2 inches). Stephen King is, and Oscar Wilde was, 1.9 metres
(6 feet 3 inches). During the eighteenth and nineteenth centuries
European Americans were, on average, the tallest people in the
world. Now, the tallest people are to be found in Herzogovina
and Montenegro, where the average height of a male is 1.86
metres (6 feet 1 1/10 inches). The second tallest are Dutchmen
at 1.85 metres. At the end of the nineteenth century the Dutch
were known as a short population. In the last 2,000 years the
shortest Londoners lived in the Victorian age. Before the twentieth century, the tallest Londoners lived in Saxon times.

Gigantism and dwarfism can produce rare and extreme
height variation, as much as 20 per cent from the average. The
tallest human we know of was the American Robert Wadlow
(1918–1940) at 2.72 metres (8 feet 11 1/10 inches).

Much of our everyday life brings us into contact with objects
that are between 1 and 10 metres in size. Nearly all the largest
living land animals are well within this range. Adult giraffes
are the tallest of the land animals, usually reaching heights
between 4.8 and 5.5 metres. The tallest known was 5.87 metres.

10–100 metres (10^1–10^2 metres)

But the *longest* extant land animal is the python. The single
longest specimen was caught in Indonesia in 1912 and measured 10.91 metres (32 feet 9 1/2 inches). Blue whales can grow
up to 30 metres long if they're allowed to live long enough.
Because of hunting most do not, and the current world population has shrunk from 200,000 to 10,000. The longest extant
animal is the bootlace worm, *Lineus longissimus*. A specimen
that turned up on the coast of Scotland at St Andrews was
around 55 metres (180 feet) long.

Land animals were larger in the past. Until recently it was thought that *Tyrannosaurus rex* was the largest of the carnivorous dinosaurs. A specimen named Sue (or, more formally, FMNH PR2081) is the largest example of *T. rex* found to date. She was 12.8 metres long and weighed perhaps 6 or 7 tonnes.[1] She is thought to have lived 67 million years ago. Fossils of *Giganotosaurus*, another kind of carnivorous dinosaur, were found in Argentina in 1993. The largest specimen so far found is 13.2 metres long. Some claim that *Spinosaurus* was the largest of all, growing to between 16 and 18 metres in length, but the original specimen, found in Egypt in 1910, was destroyed in World War II, and since then another skull is all that has been discovered.

We can assume that, whatever fossil evidence we have, it is of only a few of the many dinosaur species that ever lived. Moreover, the evidence we do have, even of those few we know of so far, is often based on just a small number of bones. There is a skeleton of *Brachiosaurus brancai* (aka *Giraffatitan*) that is unusually complete, having been pieced together from many separate finds. It stands at 12 metres tall, is 22.5 metres long, perhaps weighed up to 60 tonnes, and lived at the end of the Jurassic period, around 140 million years ago. Since the 1970s other and larger plant-eating dinosaurs have been found, though their sizes are based on incomplete, often very incomplete, skeletons. The longest and largest of all the dinosaurs is thought to have been the *Amphicoelias fragillimus* at 58 metres and 122 tonnes, but since this dinosaur has been reconstructed from a drawing of a single vertebra (the actual bone was lost) its size is speculative to say the least.

Nelson's column (including the 18-foot statue of Nelson) is about 170 feet tall. Until 2006 the column was said to be 185 feet (56.39 metres) tall. No one had thought to check.

---

[1] A *tonne* is a metric ton and equal to 1,000 kilograms, not so different from an imperial ton.

100–1,000 metres (10^2–10^3 metres)

Barely peeking into this band is the tallest tree so far discovered, a Redwood measuring 112.51 metres (369 feet 1 1/2 inches), found in 2006. Some rattan palms (genus *Daemonorops*) grow as climbers and can reach lengths of over 200 metres.

Children like to stand on top of things, and survey. Adults have retained this passion. Throughout history mankind has constructed buildings that are as high as we have been able to make them. For a brief period around 2600 BC the Red Pyramid of Snerferu in Egypt was the world's tallest man-made structure. It is thought to be the first example of a smooth-sided pyramid. Another Egyptian pyramid, the Great Pyramid of Giza, built around 2570 BC, is 146 metres (481 feet) tall and remained the tallest structure until the completion in 1311 of Lincoln Cathedral, which stands at 160 metres (525 feet). For several centuries cathedrals vied with each other for this record. Cologne Cathedral (constructed between 1248 and 1880) was the world's tallest building between 1880 and 1884. For the following five years the Washington Monument took the honour at 169 metres (555 feet), until in 1889, the year it was completed, the Eiffel Tower was measured at 300.65 metres (986 feet 4 1/2 inches) to the roof, and 312.27 metres (1,024 feet 5 inches) if the flagpole was included.

If a distinction is made between buildings and towers, 40 Wall Street was, for a very brief period, the tallest building in the world at 282.5 metres (927 feet). It was constructed in 11 months, but was exceeded in height by the Chrysler building before either building had opened. A secretly made spire added to the Chrysler building on 23 October 1929 took its height to 319 metres (1,047 feet). The dream of the American car manufacturer Walter Chrysler (1875–1940) to own the world's tallest structure lasted not much more than a year. The Empire State Building decisively took that title when it was topped out in 1931 at 381 metres (1,250 feet).

Today, the largest structure in the world is the Burj Dubai building in Dubai. The title was claimed on 12 September 2007 when it reached 555.3 metres (1,821 feet 10 inches), exceeding the height of the CN Tower in Toronto by 2 metres. On completion in 2009, the Burj Dubai aims to be over 818 metres (2,684 feet) tall.

1–10 kilometres (10^3–10^4 metres)

On normal up and down territory the horizon is a few kilometres away.[2] The horizon sets a limit on the unaided reach of our eyes across the surface of the earth, in the same way that the reach of an arm or a stride sets a limit on the physical body's reach into space.

Looking across a plain, or out to sea, and assuming you are not preternaturally tall, the absolute furthest away the horizon can be (a consequence of living on a globe this size) is about 5 kilometres (or roughly 3 miles). Obviously we can see much further if there is a prospect of distant mountains. The highest mountain is Everest at 8.848 kilometres (29,029 feet).

The deepest mine, a gold mine in South Africa called TauTona (meaning 'great lion'), is 3.6 kilometres deep.

Under the oceans, the earth's crust is between 5 and 7 kilometres thick.

10–100 kilometres (10^4–10^5 metres)

Although the tallest building in the world is less than 1 kilometre tall, the theoretical limit for a building made out of current materials is 18 kilometres (about 11 miles).

The deepest point in the Pacific Ocean is 11.034 kilometres (6.9 miles) below sea level, making the deepest oceans somewhat deeper than the tallest mountains are tall.

Many children set out to dig downwards in the hope of

---

[2] In east Kentucky, apparently, the distance to the horizon is called a 'see', as far as the eye can see.

reaching the other side of the earth. A grown-up project to drill as deeply as possible into the earth began on 24 May 1970 in the Kola Peninsula in Russia, near the Norwegian border. The deepest of several boreholes was made in 1989. Drilling stopped in 1992 when it became clear that the underground temperatures of 300°C made it impossible to prevent the drill bits from melting. The deepest of these holes and the deepest hole that mankind has ever made was 12.262 kilometres (7.62 miles) deep.

Under the continents, the earth's crust is, on average, 34 kilometres thick, and up to 80 or 90 kilometres in some parts.

The highest clouds (known as noctilucent clouds) are silvery-blue and usually form in the summer months about 80 kilometres above the poles, though in recent years their number has increased and they have been seen as far south as Utah.

In the United States, an astronaut is anyone who has travelled more than 80.5 kilometres above the surface of the earth.

The atmosphere of the earth has no boundary. It just endlessly thins. Three-quarters of the mass of the atmosphere, however, is held within 11 kilometres of the surface of the earth. For practical purposes, the edge of the atmosphere is defined by the Kármán line, named after the Hungarian-born American engineer Theodore von Kármán (1881–1963), who discovered that at around 100 kilometres it becomes difficult to achieve aerodynamic lift.

When the earth passes through an area of dust and small rocks – usually the debris left behind from when a comet has passed near the sun – some of this material may enter the earth's upper atmosphere. The friction caused by the impact we see as shooting stars. The Perseid meteor shower, also known as the tears of St Lawrence, has been observed in the northern hemisphere every August for over 2,000 years. Each year, hundreds of tons of fine dust particles simply float down from outer space to the surface to the earth. Larger lumps of matter that reach the surface of the earth are called meteorites.

100–1,000 kilometres (10^5–10^6 metres)

Military satellites orbiting the earth at 500 kilometres can pick out objects on the ground 20 centimetres long.

In 1990 the Hubble telescope was launched. It orbits at a height of 600 kilometres. The telescope's effectiveness was initially compromised when it was discovered that the main mirror had been ground incorrectly. A remarkable correction made in space in 1993 restored the telescope's intended capability. Overnight, information collected by this telescope doubled the estimated number of stars in our galaxy, the Milky Way.

Several satellites whose days of service have come to an end have been blown up. On 11 January 2007, China exploded its Fengyun 1C weather satellite into at least 2,400 pieces larger than an orange. It will take centuries for all the pieces to fall to earth. Some satellites have had to be moved into new orbits as a result. It is estimated that there are at least 18,500 pieces of man-made debris larger than 10 centimetres across and 600,000 pieces bigger than 1 centimetre orbiting the earth below 1,000 kilometres.

There are presently 417 satellites orbiting the earth at heights ranging between 160 and 2,000 kilometres above ground, that is, straddling this order of magnitude and the next. Satellites in these orbits are called Low Earth Orbit satellites (LEOs). The International Space Station (ISS) that is currently being assembled in space is a LEO. It moves from between 319.6 and 346.9 kilometres above the earth, and orbits 15.77 times a day. It can be seen from earth with the naked eye.

In 1948, Fred Hoyle predicted that the first photograph of the earth seen from the outside would be 'a new idea as powerful as any in history'. The first such image was taken from Apollo 8 in December 1968 at an orbit of between 181.5 and 191.3 kilometres. This photograph, called 'Earthrise', is indeed said to have made a significant impact on the philosophy of environmentalism, a movement that began to take off in the 1970s. A television broadcast from Apollo 8 made on Christmas Eve of the crew reading from the Book of Genesis was at the time the most watched TV programme in history.

1,000–10,000 kilometres (10^6–10^7 metres)

There are currently 47 man-made satellites orbiting the earth between 2,000 and 35,800 kilometres above ground, that is, straddling this order of magnitude and the next. They are called Medium Earth Orbit satellites (MEOs). The most famous MEO is probably the Telstar satellite launched in 1962. It was the first communications satellite. Its first broadcast was meant to be a televised message from President Kennedy, but since he was not prepared the first broadcast turned out to be part of a major league baseball game between the Philadelphia Phillies and the Chicago Cubs. Telstar 1 ceased broadcasting in 1963 but is still in orbit.

The Great Wall of China is about 4,000 kilometres long. The distance from the surface of the earth to its centre is 6,370 kilometres. The longest river on earth is the Nile, which flows for 6,695 kilometres.

10,000–100,000 kilometres (10^7–10^8 metres)

Another consequence of living on a globe this size is that we are never more than 19,000 kilometres from home. We can be no further away from home without going around in circles or returning by another route.

The so-called 'Blue Marble' photograph of the earth was taken from a height of 28,000 kilometres by Apollo 17 in 1972, another photograph that spurred on the environmental movement.

There are currently 351 man-made satellites in orbit at or above 35,786 kilometres. Such satellites are called High Earth Orbit satellites (HEOs).

100,000–1,000,000 kilometres (10^8–10^9 metres)

Vela 1A is an example of an HEO. It was launched in 1963, three days after the Test Ban Treaty was signed, and was designed to detect nuclear explosions from space. It orbits at a little over 100,000 kilometres above the earth.

There is no alternative now, but to leave the earth behind, to let go of this slight obsession with nice distinctions between the size of earthly things. It's time to go beyond the atmosphere, beyond the man-made satellites, and to look out across space to the next nearest sizeable object.

The moon, the earth's natural satellite, is on average 384,399 kilometres away, about a quarter of a million miles. At its furthest reach it is 405,696 kilometres away from earth, and at its closest, 363,104 kilometres. The moon is illuminated, as everything in the solar system is, by the sun. The sun is a star and only stars shine. The moon appears to be the next brightest object after the sun, but what we see (and call moonlight) is merely the reflected light of the sun. The moon appears to us to be the brightest object in the night sky. Even during a full moon, moonlight is 500,000 times less intense than sunlight: too weak to show the world in colour. During clear nights when the moon is a sliver in the sky, it is possible to see the sunlight that illuminates the earth reflected back on the moon. On such nights, alongside the bright sliver, a dim impression of the whole moon can be made out. The bright sliver is the moon illuminated by the sun, and the dim whole is the moon illuminated by earthshine. Leonardo da Vinci (1452–1519) was the first to account for this effect correctly.

1–10 million kilometres (10^9–10^{10} metres)

Space is called space for a reason. You would be hard pushed to find any sizeable solid object in this region excepting the odd meteorite, or rare passing asteroid. Space, however, is far from empty. Radiation and atoms are everywhere.

10–100 million kilometres (10^{10}–10^{11} metres)

After the moon, Venus is the next closest sizeable object that we could come across in our walk out from the earth. At its closest Venus is 40 million kilometres away. It appears to be the second brightest object in the night sky.

100–1000 million kilometres (10^{11}–10^{12} metres)

The sun is on average 150 million kilometres away. The sun's average distance from the earth is called an astronomical unit (AU), a handy unit of measurement astronomers use to navigate their way around the solar system and its environs.

The sun is made up of more than 99.9 per cent of all the matter in the solar system, and so the gravitational influence the planets have on the sun is insignificant compared with the gravitational effect of the sun on the planets. We tend to say that the planets orbit the sun, but it would be more accurate to say that they orbit about a common gravitational centre. Given the overwhelming mass of the sun, that gravitational centre is, however, very close to being the centre of the sun itself.

Although we might have stumbled across Venus when we made our previous step through space, the orbit of Venus also sweeps it out into this more distant region. On average Venus is a little over 100 million kilometres away. Mars and Mercury make similar sweeps between the previous zone and this one, depending on what side of the sun they lie relative to earth.

Here we might pause and wonder whether measuring the distance of far-flung objects from the earth is a sensible thing to be doing. The sun declares itself the physical centre of the solar system simply by virtue of the fact that it contains nearly all the mass of that system. Overall the sun is not very dense – its average density is one and a half times that of water – which means that given its great mass it is also large (some 1.4 million kilometres across).

We could continue to measure the universe from the earth, but we can already begin to see that such an arrangement looks contrived. The universe turns out to be a place where stuff is in motion, and when we take notice of stuff and motion it is clear that the planets defer to the sun. Judged by mass and motion, we see Mercury making the closest orbit of the sun, followed by Venus, the third rock that is the earth, and then Mars beyond. Collectively, these are the terrestrial planets. With the sun at the physical centre of the solar system, the physical relationship between the various bodies that make up that system is made apparent.

The asteroid belt is a band of rocky rubbish left over from when the terrestrial planets were first made. It separates the planets that have a visible surface from the gaseous planets, and envelops a region that stretches from about 270 million to 675 million kilometres away from the sun (or 1.8 to 4.5 AU). Asteroids range in size from a grain of dust to the minor planet Ceres which is 950 kilometres in diameter. There are three other major lumps of rock each about 400 kilometres wide, and together these four bodies make up most of the mass of the belt. The Sloan Digital Sky Survey that began to survey the skies in the year 2000 has so far detected 600,000 asteroids. It is thought the survey will have photographed a million of them by 2017.

The orbit of some asteroids sometimes crosses the orbital path of earth, occasionally at the same moment. On average, collisions with asteroids about 5 kilometres across happen every 10 million years, with asteroids 1 kilometre across every million years, and with those 50 metres across every 1,000 years or so. An asteroid between 5 and 10 metres across explodes in the upper atmosphere every year with the equivalent force of the atomic bomb that was dropped on Hiroshima. An asteroid 50 metres wide (perhaps wider) exploded over the Tungaska river valley in Siberia in 1908 and laid to waste 2,000 square kilometres of forest. The 300-metre-long asteroid 4581 Asclepius missed the earth by 700,000 kilometres on 23 March 1989, that is to say it passed exactly where the earth had been six hours earlier. If it had struck, it is estimated that the explosion would have had the force of a Hiroshima-sized bomb detonated every second for 50 days. Such near misses are often only realised after the event. So far, 800 so-called PHAs (potentially hazardous asteroids) have been detected. There are thought to be another 200. By congressional mandate, NASA is cataloguing all near earth objects (any object that cuts across the earth's orbit, not just asteroids) wider than a kilometre. The asteroid 1940DA, which is 1 kilometre wide, might collide with the earth on 16 March 2880.

The first gaseous planet we arrive at is also the largest. Jupiter is more than twice the mass of all the other planets added together. It is on average a little over 778 million kilometres from the sun (or about 5 AU). Jupiter appears to us to be the third brightest body in the night sky after the moon and Venus.

1–10 billion kilometres (10^{12}–10^{13} metres)

Saturn is the sixth planet furthest out from the sun and the second largest after Jupiter. It is, on average, a little more than 1.4 billion[3] kilometres distant from the sun. We are now so far out into space that our measurements still lie in the same band whether we make them from earth or from the sun. Measured from *earth*, Jupiter is, on average, a little less than 1.3 billion kilometres away.

Uranus is the seventh furthest at 2.8 billion kilometres distant. It is the third largest planet by size and fourth largest by mass. Uranus is also the first planet to be discovered in modern

---

[3] A billion is a thousand million. Astronomers get used to separating the vast from the vaster. A change of unit of measurement often helps. In everyday life we often use a million to mean a 'lot', or we describe a rare occurrence as one in a million. Scientists, particularly cosmologists, tend to use a billion in much the same way. It is often their shorthand for a lot when more precise information is wanting. You may be surprised how often a billion is the answer to cosmological questions.

times. On 13 March 1781, the German-born English astronomer William Herschel (1738–1822) noticed that what had
hitherto been identified as a star must be some other sort of
heavenly body. At first he thought it was a comet but by 1783
it was clear that he had discovered a new planet. The solar
system expanded for the first time in the modern era. For his
discovery, Herschel was granted an annual pension of £200
by King George III.

Neptune is the eighth planet out from the sun, and the
fourth largest by size and third largest by mass. It is around
4.5 billion kilometres away.

We see because the light of the sun shines on things, but
we can also 'see' through the gravitational influence of one
body on another. In the early decades of the nineteenth century
it was noticed that the orbit of Uranus is perturbed in such a
way that the presence of a massive and hitherto unaccounted
body was indicated. It was this prediction that led to the
discovery of Neptune. Science often works this way: a prediction that such and such an entity must exist directs the scientist where to look. If the prediction is true, then the scientist
stands a better chance of actually bringing the entity out into
the light of day.

That light and gravity are the two means of finding out
what is out there in space might suggest that there is some
connection between them. Discovering the nature of that
connection is the story of modern physics, and a theme of
this book.

Uranus and Neptune are distinguished from the other gas
planets for having greater proportions of ice (it is cold out
here) and are for this reason known as the ice giants.

All objects in the solar system that are further out than the
orbit of Neptune are called trans-Neptunian objects.

Pluto, once thought of as the smallest planet in the solar
system, is now described as a dwarf planet. Pluto has an eccentric orbit that brings it closer to the sun than Neptune but also
further out than the furthest reach of Neptune's orbit. Discovered in 1930, Pluto ceased to be a planet in August 2006, when

it was downgraded to a dwarf planet[4] and given the number 134340, the astronomical equivalent, one can't help but feel, of being sent to the back of the class. At the time of writing, the entry on Pluto in the online resource Wikipedia has been locked because of vandalism. Presumably, attempts have been made to reclaim its planetary status. Pluto's status was undermined by the discovery of Eris, another dwarf planet and trans-Neptunian object. It was discovered in 2005, and is larger than Pluto in both diameter and mass. From 11 June 2008, Pluto, Eris, and other trans-Neptunian dwarf planets are called plutoids. To date, Makemake is the only other dwarf planet that is also a plutoid; but there are at least another 41 trans-Neptunian objects that may eventually make the grade. Makemake was discovered in 2005 and has a diameter about three-quarters that of Pluto.

Many trans-Neptunian objects are held in a region called the Kuiper belt, which extends beyond Neptune to about 7.5 billion kilometres (or 50 AU) away. The Kuiper belt is home to about 35,000 small solar system bodies wider than 100 kilometres. All the comets with short-period orbits live here, meaning that these comets return for a subsequent sighting after relatively brief periods of time. Halley's comet is a short-period comet that returns every 75 or 76 years. Comets are made of ice and dust. Some of them, such as Halley's comet, have eccentric orbits that take them close to the sun. At such times, removed from the cold depths of the solar system, some of the ice will be melted by the heat of the sun and be seen from earth as a tail. Although we talk of a comet's tail, the 'tail' is vapour blown away from the comet by a stream of particles emanating from the sun called the solar wind. So the

---

[4] Planets are those astronomical bodies in the solar system that orbit the sun and are massive enough to have been made into rounded objects by the force of their own gravity. Our moon would count as a planet if it were not under the gravitational influence of the earth. A dwarf planet is massive enough to be rounded by gravity but is not massive enough to have cleared the area of small irregular bodies smaller than planets that are called, plainly enough, small solar system bodies.

tail points away from the sun regardless of whether the comet is approaching or receding from it. Halley's comet is 35 AU away from the sun when it is at the furthest point of its eccentric orbit, and only 0.6 AU away at its closest.

Objects in the Kuiper belt are thought to have remained unchanged since the early days of the life of the solar system making these comets prized objects of investigation. Wild 2, a comet recently visited by a NASA craft, originates from the Kuiper belt but has been moved into a closer orbit by the gravity of some large body, at a time when the solar system was younger and more volatile, making it of particular interest since it is now close enough to be visited and analysed. Its composition has much to tell us about the early conditions of the solar system.

10–100 billion kilometres (10^{13}–10^{14} metres)

The journey beyond the Kuiper belt takes us into another region devoid of large physical objects. Discovered in November 2003 and three times further out than Pluto is tiny Sedna (about two-thirds the size of Pluto), which may have come from the Kuiper belt or, because its orbit is so elliptical, from the Oort cloud (a place we have not arrived at yet). Sedna is 13.5 billion kilometres away, and the furthest object in the solar system that has been observed. Not surprisingly, given its distance from the sun, it is also, at −240°C, the coldest object in the solar system. Sedna, named after the Inuit goddess of the sea, who lives in the deepest parts of the Arctic ocean, is currently only 70 years away from being at its closest to earth on its 11,487 year orbit.

The solar wind blows away the interstellar gas (hydrogen and helium left over from the early days of the life of the universe) to make a vast bubble with a radius longer than the distance to Sedna. This bubble is sometimes taken as a definition of the limits of the solar system. The outer edge, where the wind is not strong enough to blow the gas away, is turbulent and is called the heliopause (after Helios the Greek god of the sun).

The furthest and fastest-moving man-made object in the universe is now approaching this turbulent region. The 722 kilogram space probe named Voyager I began to move out of the solar system in 2004. It is a little further out than Sedna, 14.4 billion kilometres away, though the 'little' is relative. Voyager 1 is already as far beyond Sedna as six times the distance between the earth and the sun. Voyager 1 left earth in 1977 to visit Jupiter and Saturn.

100–1,000 billion kilometres (10^{14}–10^{15} metres)

We pass through a further realm seemingly lacking in large objects; or if not empty, whatever large objects might lie here are as yet unseen. The sun is the only lamp in the solar system. Only what falls into its pool of light is illuminated. To find other sources of light we must look further out, to other stars. The sun is a special name we give to our local star, but it is just one of many stars, those bodies in the universe that shine.

We see other suns (and collections of suns called galaxies) more clearly than we can see the edges of our own solar system. In some ways we know less about the solar system, particularly at its furthest reaches, than we know about the universe at large. Part of the problem is how to throw light on this region of the solar system when the only light comes from the sun. At this distance the light fails. There may be many objects out here, even of considerable size, that we cannot see because they do not reflect enough of the sun's light to be visible, and are also too distant to be detected gravitationally.

It has been suggested that the highly eccentric orbit of Sedna is evidence that the sun has a dim companion. When stars are formed they tend mostly to be formed as pairs of stars, as binary systems, or in groups of three. The sun would be unusual (though not unique) for being solitary. But if the sun is indeed part of a binary pair, then how do we account for the fact that its companion star has not yet been seen? It has been conjectured that we do see it: that it occasionally nudges

some of the distant comets into closer orbit. This slight gravitational influence is not yet sufficient evidence to elevate this
conjecture into a theory.

1,000–10,000 billion kilometres (10^{15}–10^{16} metres)

At the furthest reaches of the solar system, 50,000 times the
distance from the earth to the sun, a thousand times further
out than the furthest planets, and at the effective limit of the
sun's gravitational power, is the Oort cloud, or so it is conjectured. There is no direct evidence for its existence, but in 1950
the Dutch astronomer Jan Oort noticed that there are no
comets that appear to come from interstellar space, that is,
with orbits beyond the gravitational influence of the sun.

The Oort cloud is thought to be the home of all comets with
long-period orbits. Some of these orbits may take millions of
years to complete. There are perhaps as many as a billion or
even a trillion (a thousand billion) comets in the cloud. It is
called a cloud, not just because there are so many of them, but
because the many objects orbit at every conceivable angle. In
the Kuiper belt all the comets orbit in the same plane. Curiously, the Oort cloud may house objects that were once closer
to the sun than the comets to be found in the much nearer
Kuiper belt. The Oort cloud holds the light (and now unseen)
bodies that the large gas planets threw into more distant orbits.
The faintest effects of the combined gravitational field of the
sun and the planets just barely holds them in at these great
distances. In the last 300 years, 500 long-period comets have
been identified.

Like Kuiper objects, the bodies that make up the Oort cloud
have remained unchanged since the solar system formed.

Handily, the furthest objects in the solar system are a little
under a light-year away (10^{16} metres is close to the distance
that light travels in a year). Continuing at its current rate of
0.006 per cent of the speed of light, Voyager 1 will be this far
out in about a thousand years from now.

If the sun were the only massive object in the universe, we would observe its gravitational influence extending infinitely and ever more weakly; but in practical terms, at these distances and as we begin to move into the gravitational ambit of other massive bodies, the sun's power is at an end. The light-year is a familiar unit that measures distance, even though it sounds as if it should be a measurement of time. The romance of the word light-year hints at a connection between time and space, a connection that will become more apparent as we progress further. When we look out to this region of the solar system, which as it happens we cannot see, we are also looking back a year in time.

1–10 light-years or 10,000–100,000 billion kilometres (10^{16}–10^{17} metres)

The next large object that we come across is our nearest neighbouring star, Proxima Centauri, which is a little over four light-years away. It cannot be seen from earth with the naked eye. Like most dim stars less than half the mass of the sun, Proxima Centauri belongs to a group of stars called red dwarfs. It was first observed from earth in 1915. A little beyond it at 4.37 light-years are the stars Alpha Centauri A and B: A is a little larger and brighter than the sun, and B is a little smaller and dimmer than the sun. Together they can be seen from earth with the naked eye. The two stars are, however, easily resolved even with the smallest of telescopes. Alpha Centauri's binary nature has been known about for over 200 years. It is now thought that Proxima Centauri is part of the same star system. The next nearest stars are Barnard's star (5.96 light-years away), Wolf 359 (7.78 light-years away), Lalande 211 85 (8.29 light-years away), Sirius A and B (8.58 light-years away), Lutyen 726-8 A and B (8.78 light-years away) and Ross 154 (9.64 light-years away). The average distance between stars in our galaxy is about 3.3 light-years, somewhat less but not so different, if we allow ourselves to be cavalier about the odd light-year, from the distance between our sun and it nearest neighbour.

10–100 light-years (10^{17}–10^{18} metres)

The next nearest star systems measured in light-years is a list of numbers that reads: 10.32, 10.52, 10.74, 10.92, 11.27, 11.40, and so on. In the bowl of space that surrounds the sun 16.31 light-years in every direction, 50 stellar systems have been found so far. The list isn't definitive, and there are doubtless other nearby stars that have not yet been discovered. A research programme aiming to catalogue all the nearest stellar systems lists 2,029 systems in the bowl that reaches out 32.6 light-years.

These stars are in various stages of their life cycles. A star more than half the mass of our sun will, towards the end of its active life, enter a phase where it swells up to many times the size of its core. Its outer layers are dramatically dispersed making the star look enormous. During this phase it is called a red giant. Our own sun will not do this for another 5 billion years. Arcturus is a red giant 36.7 light-years away. Though it is less than one and a half times the mass of the sun, its power output is about 180 times that of the sun, making it appear to us to be the third-brightest star in the sky.

51 Pegasi is a star some 50.1 light-years away. In human history it is notable for being the first solar system other than our own that we have looked on. 51 Pegasi is a sun-like star (though somewhat older) with at least one other planet orbiting it. Since this planet's discovery in 1995, around 300 other so-called exoplanets have been found. It is thought that about 1 in 14 stars is the centre of a planetary system. Upsilon Andromedae is a triple star system 44 light-years away that has multiple planets in orbit around the main star.

Aldebaran (aka Alpha Tauri), is another red giant 65 light-years away. It is 38 times the diameter of the sun, and 150 times brighter. It is seen from earth as the fourteenth brightest star in the sky. Gacrux (aka Gamma Crucis) is a red giant 88 light-years away.

In order to arrive at these many large objects, we would have to set out in many different directions. We distinguish these different possible journeys by reference to the constellations that we see in the night sky from earth. The constellations are arbitrary collections of stars that have been differently named and differently conceived in different cultures and at different times in history. Today, for example, we say that the 51 Pegasi is in Pegasus, by which we mean that we might arrive at the 51 Pegasi if we set off in the direction of that patch of the sky where we once drew out among the stars the shape of a winged horse. Aldebaran is in Taurus, meaning that it lies in that general direction where ancients saw a bull. The constellations are compasses, and like all compasses they tell us nothing about the distance to the objects being pointed at. The constellations are themselves made up of a number of bright objects, each of which may be at quite a different distance from any other. They are strange compasses, made out of the very things they are pointing out. Strange, too, because they are protean: over longer periods of time than human history has experienced, the patterns made by these stars change. From our human perspective, the stars do not appear to move. But they do move, and only appear stationary because they are so far away. We move to a different beat.

100–1,000 light-years (10^{18}–10^{19} metres)

Betelgeuse (pronounced *beetlejuice*) is a red supergiant with a peculiar name. A red supergiant is like a red giant, just bigger. Betelgeuse is fifteen times as massive as our sun but 40 million times greater in volume. It is about 427 light-years away, and has a diameter four times the distance between the earth and sun (4 AU), making it one of the largest stars in the sky. It is also the ninth brightest. Betelgeuse comes from an Arab word. What the word might be is disputed. It could be the word for a black sheep with a white spot in the middle of its body. It could be *yad al-jawzā*, meaning hand of the central one, which was translated into Latin as Bedalgeuze. In the Renaissance, the word was thought to have been *bait al-jawzā*, or armpit of the central one, which was retranslated into Latin as Betelgeuse.

Eventually the material around the core of the star will disperse. It is possible that the core will explode, that is, become a supernova. Opinion is divided as to whether this will happen or not. If it does, Betelgeuse would become as luminescent as the moon for several months. Some sources say that Betelgeuse may already have exploded: we just don't know it yet. The light from that explosion will take 427 years to get here.

1,000–10,000 light-years (10^{19}–10^{20} metres)

The Orion Nebula, or M42, is a thin cloud of dust and gas 1,500 light-years away, where thousands of stars are being formed out of the debris left from the explosion – as supernovae – of previous generations of stars. It is 30 light-years across and the closest star-forming region in our galaxy. Nebula is derived from the Latin word for mist. M42 is Messier object 42. Charles Messier was an eighteeth-century French astronomer, and cataloguer of the night sky. His list of astronomical bodies labelled M1 to M103, together with seven later additions, is still used today.

Large supergiants are called hypergiants. VY Canis Majoris is a hypergiant 5,000 light-years away. It is over twice as wide as Betelgeuse (and between 1,800 and 2,100 times as wide as the sun), making it the largest (but not most massive) of all the known stars.

M1 is the Crab Nebula and it is 6,300 light-years away. Unlike the Orion Nebula, it is a cloud formed from the explosion of a single star (what Betelgeuse may become). In AD 1054, the year it was first observed by Chinese and Arab astronomers, it looked like a star brighter than any others in the sky. At the moment of explosion a supernova can outshine its host galaxy for several weeks and emit more energy than the sun will in its 10-billion-year lifetime. Today, this supernova is seen as a cloud 6 light-years across. It was named the Crab Nebula by the third Earl of Rosse, William Parsons. In 1844 he drew what looked like a crab when he observed the nebula through his telescope. When he looked again in 1848, through a larger telescope, he realised that it did not in fact look like a crab, but by then the name had stuck. In 1968, the neutron star at its centre was discovered, the remnant of the original star collapsed into a very dense body just 30 kilometres across. The neutron star is mostly made of densely packed neutrons, a subatomic particle found in the nucleus of most atoms. A lump of this matter the size of a sugar cube weighs 100 million tonnes. This particular neutron star spins on its axis 30 times a second and emits radiation through every part of the spectrum from radio waves to gamma radiation. A spinning neutron star is called a pulsar and the neutron star at the heart of the Crab Nebula was the first pulsar to be observed. Pulsars possess the strongest magnetic fields in the universe, about 100 billion times stronger than the earth's.

The Boomerang Nebula is one of the most peculiar nebulae so far discovered. It lies 5,000 light-years away and, at −272°C, is the coldest place in the universe, just one degree warmer than the coldest possible temperature. Since temperature is a measure of the average motion of a collection of molecules, the coldest temperature is reached when molecules possess the least motion. At absolute zero (−273°C), a theoretical temperature that cannot actually be reached, molecules would no longer be moving. According to quantum physics, it is impossible for molecules to be completely at rest.

Why the gas cloud around the star at the centre of the Boomerang Nebula is so cold is not entirely understood. There appears to be some unique way in which carbon monoxide is being expelled from the star, as a very cold wind that is reducing the surrounding temperature. The nebula was discovered in 1998 by the Hubble space telescope.

10,000–100,000 **light-years (10^{20}–10^{21} metres)**

SN1604 or Kepler's supernova was suddenly visible to several earthbound observers on 9 October 1604. It is named for the great German astronomer Johannes Kepler (1571–1630), who was among the first observers. Kepler's supernova is 13,000 light-years away and is the most recently observed supernova in our galaxy. At the time, Kepler's supernova shone almost as brightly as Venus for several weeks. The bright light that those first observers recorded had travelled for 13,000 years to reach them.[5] The light from Venus, however, only takes a few minutes to reach us. We look out into the night sky and record an impression of what is out there and we are inclined to think that that is how it is *now*, at this moment. Yet that *now* is made up of many *nows*, overlaid to make what we take to be a record of some event in the life of the universe rather than some subjective experience. Where *now* is in the universe is no more apparent than where its centre is.

Canis Major dwarf galaxy (found by heading in the direction of the constellation named Canis Major) contains a billion stars (small for a galaxy) and is our nearest neighbouring galaxy. It is a satellite galaxy held within the confines of our own much larger galaxy (as a planet is held by the sun in the solar system). It was discovered as recently as November 2003. Sometimes it is hard to see what is, in astronomical terms, under our noses. It can be difficult to work out the shape and content of our own galaxy (which we have named the Milky Way) given that we are in it, and so with no outside vantage point. A similar but more intractable problem applies to the universe as a whole, from which there is no outside vantage point, except perhaps in human imagination.

Canis Major dwarf lies some 42,000 light-years from the gravitational centre of the Milky Way, but is 25,000 light-years

---

[5] Light from Kepler's supernova has taken several hundred years longer to reach us, and today we see that the source is no longer so bright. It was only as bright as when Kepler and others saw it for a few weeks.

away from the solar system. That it is our nearest neighbouring galaxy is only of interest to us.

At this point, we can begin to see that it makes more sense to make our measurements from the gravitational centre of our galaxy rather than from the sun (the gravitational centre of the solar system). A material description of reality sees the universe as an arrangement of massive objects moving about each other. The physical evidence that the stars in our galaxy move about its gravitational centre is the simple and overwhelming reason we change perspective. And from this new perspective it is clear that the centre of the galaxy is more privileged than the centre of the solar system. We *could* describe the contents of the Milky Way, and the position of neighbouring galaxies, from the vantage point of our solar system, but the stuff of the universe is more elegantly described as the motion about the common gravitational centres of ever-larger structures. The planets revolve around the sun, the sun moves around the gravitational centre of the galaxy. Our journey into space, we begin to see, is the search for these ever larger structures. In a material description of the world material dominates.

Galaxies are typically 10,000 light-years across, though the Milky Way is between eight and ten times bigger than the average. Dwarf galaxies that gravitationally attach themselves to larger galaxies, as Canis Major dwarf has done to ours, might be only tens of light-years across.

The solar system is thought to be 26,000 light-years from the centre of the Milky Way. The estimate of this distance has changed substantially in recent years, revised downwards from 35,000 light-years. Not only are we not at the centre of the solar system, neither are we at the centre of the galaxy. In fact we are slightly closer to the satellite galaxy Canis Major dwarf than we are to the centre of our own galaxy.

Given that there is a black hole called Sagittarius A at the centre of our galaxy it is just as well that we are some distance away. The mystery of the existence of black holes has turned them into curiously romantic entities. Black holes are concentrations of matter so dense that even light cannot escape them:

they are what a neutron star would become if it were denser. It is well known that a rocket must attain a certain velocity if it is to escape the earth's gravitational pull, that is, reach the earth's escape velocity. Light travels so fast – in fact at the fastest speed possible according to the laws of nature as we currently understand them – that a body must be incredibly massive (and hence its gravitational field incredibly strong) if its escape velocity is to exceed the velocity of light. Black holes are such bodies. Light cannot escape them.

Sagittarius A, whose existence as a black hole was generally agreed upon in 1996, is thought to be 3 million times more massive than the sun. It is current thinking that a black hole may lie at the centre of most if not all galaxies.

Estimates of the number of stars in the Milky Way vary from between 200 and 400 billion stars. If most of them are smaller than the sun, as is widely believed, then the higher estimate seems likely. Our galaxy is a rotating flattened disc 100,000 light-years across (and on average 1,000 light-years thick), made of spirals of stars and dust and gas surrounded by a larger and less-densely populated sphere of stars called a halo. The spiral arms trace out the same spiral shape that we see in nautilus shells and in cyclones. This is where the young, hot, bright stars are to be found (and is where we are). There are four major spiral arms: Perseus, Sagittarius (no connection to the black hole Sagittarius A), Centaurus and Cygnus. The solar system lies on a small arm called Orion between the outer arm Perseus and the inner arm Sagittarius. Orion may even be a spur of the Perseus arm. Within these spiral arms of gas – thinly populated with young stars like our sun, called Population I stars – most of the current star-forming activity of the Milky Way takes place.

At the centre of this disc is a bulge densely populated with old stars some 10,000 light-years wide and 3,000 light-years thick. Lying across that central bulge is a bar of stars about 27,000 light-years wide. This central bar was discovered in 2005 and also mostly consists of old stars, either red giants or those small dim stars called red dwarfs.

100,000–1,000,000 light-years (10^{21}–10^{22} metres)

Enveloping the spiral disc is a huge sphere, or halo, 200,000 or more light-years across. The halo is thinly populated with more old stars, some of which have gathered together into globular-shaped clusters. No new star formation takes place in this region. There are about 150 globular clusters in the Milky Way. It is thought that there are more to be discovered, perhaps between 10 and 20 more. Each globular cluster contains hundreds of thousands of stars. The clusters orbit the gravitational centre of the galaxy at huge distances, over 100,000 light-years away.

There are several dwarf galaxies like Canis Major dwarf that have become bound to our own galaxy. The largest of them is the Large Magellanic Cloud (LMC), visible only in the southern hemisphere, and named for the Portuguese explorer Ferdinand Magellan (1480–1521). He observed the LMC on his famous voyage of 1519, the first time a European crossed the Pacific by first heading westwards. The LMC had also been observed by the Italian explorer Amerigo Vespucci (1454–1512) during a voyage made a few years earlier; and hundreds of years before that, the Persian astronomer 'Abd al-Rahman al-Sufi writes of it in his *Book of the Fixed Stars* (*c.* AD 964) where he calls it the White Ox. The LMC is 179,000 light-years away, and is made up of about 10 billion stars. It is about half the diameter of an average galaxy (or a twentieth of the diameter of our atypically large galaxy).

Though bound to our galaxy, the LMC is fated to be consumed by Andromeda, the nearest neighbouring galaxy that is gravitationally independent of us. Canis Major dwarf galaxy, on the other hand, is in the process of being encompassed by the Milky Way.

On 24 February 1987 a supernova exploded in the LMC. It was the nearest observed supernova since Kepler's of 1604.

There is also a Small Magellanic Cloud, another dwarf galaxy. It contains less than a billion stars and is 210,000 light years away.

Perhaps now is the moment to observe that the universe we have described so far is one of local cultural interest. Aliens would have their own intergalactic tourist sights, on which the LMC, for example, would almost certainly not feature.

1–10 million light-years (10^{22}–10^{23} metres)

Barnard's galaxy, yet another dwarf galaxy gravitationally bound to our own, is 1.6 million light-years away. It is 200 light-years across. One of the easiest galaxies to see through a telescope, Barnard's galaxy was discovered in 1881, though it was not at that time recognised as a galaxy. At that time, and up until the 1920s, it was thought that there was only the one galaxy: that the whole of the universe was the Milky Way.

Two and a half million light-years away, we find Andromeda (M31), our nearest large neighbouring galaxy. It is twice the size of even our unusually large galaxy. Both Andromeda and the Milky Way have 14 known satellite galaxies. Andromeda is a spiral galaxy like ours. Not all galaxies are made up of spiral arms. Some are elliptical, measured on a scale E0 to E8, with E0 galaxies the most circular and E8 the most elliptical. Any galaxy that can't be characterised as a spiral or an elliptical galaxy is called a peculiar galaxy. It is not yet known how the older elliptical and peculiar galaxies are formed, perhaps as a result of the collision of spiral galaxies.

Andromeda is, neatly, the furthest object we can see with the naked eye. It looks like a dim star.

10–100 million light-years (10^{23}–10^{24} metres)

Gravity pulls galaxies together, just as, at smaller sizes, it pulls the planets around the sun or apples to the earth. The Milky Way belongs to a small cluster of gravitationally held galaxies called the Local Group. It is 10 million light-years across, and made up of about 40 galaxies, some of them very small like the satellite galaxies Canis Major dwarf, Barnard's, and Sagittarius dwarf elliptical. By far the largest galaxies in the group are our own galaxy and Andromeda, then, trailing some distance behind, is a galaxy called the Triangulum.

Though the Milky Way and Andromeda are said to be gravitationally independent, it is all a matter of degree. The fate of dwarf galaxies appears to be sealed: it is predicted that they will be swallowed and torn apart by their hosts or by larger neighbouring galaxies. The fate of Andromeda and the Milky Way is similarly predicted, but we have to look at a longer time span. These two massive galaxies, revolving around their common gravitational centre, are like two wrestlers facing each other off. In 3 billion years' time they will pass through each other, the beginning of a process that may then take several more billion years before the black holes at their centres come together to make a super-sized black hole at the centre of a super-sized galaxy. The combined galaxy may ultimately change shape and become an elliptical galaxy.

The nearest cluster of galaxies neighbouring our own is the Virgo cluster, some 60 million light-years from the centre of the Local Group. It is so large – comprising perhaps as many as 2,500 galaxies – that it pulls the Local Group gravitationally.

For us, making this particular journey, the centre of the Milky Way has made way for the centre of the Local Group, which, in turn, makes way to the gravitational centre between the Local Group and the Virgo cluster. The hunt for the centre of the universe moves on.

100–1,000 million light-years (10^{24}–10^{25} metres)

Clusters of galaxies, like the Local Group and the Virgo cluster, clump together to make superclusters: clusters of clusters of galaxies. The Local Group belongs to the Virgo supercluster (not to be confused with the Virgo cluster) of galaxies, and is made up of 2,500 bright galaxies; that is, we can see them, and there might be more. There are about a hundred clusters of galaxies in Virgo supercluster dotted about a region 200 million light-years across. The Local Group is located at an outer edge. Because of its size, and hence its gravitational influence, the Virgo cluster is close to its centre. It is becoming very clear that we are determinedly not at the centre of anything, unless the 'we' becomes ever more inclusive.

Our nearest neighbouring supercluster is the Hydra–Centaurus supercluster, between 100 and 200 million light-years away. The Coma supercluster, another nearby supercluster, is around 300 million light-years away. There are thought to be 10 million super-clusters in the universe and hardly any galaxies in between them.

The Coma supercluster lies in the centre of the second-largest structure so far discovered in the universe, the Great Wall, discovered in 1989. It's a strand of super-clustered galaxies about 200 million light-years away and thought to be about 600 million light-years long, though it could be longer. It is 300 million light-years wide but only 15 million light-years thick.

1–10 billion light years (10^{25}–10^{26} metres)

The largest object so far discovered in the universe is called the Sloan Great Wall, which was discovered on 20 October 2003 from data collected by the Sloan Digital Sky Survey. It is a filament of superclusters and clusters of galaxies about 1 billion light-years away and almost 1.5 billion light-years long. It would take 250,000,000,000,000,000 copies of the Great Wall of China placed end to end to cover this distance. There is no general agreement that the Sloan Great Wall is a true structure, given that its parts are not gravitationally bound to each other.

The Sloan[6] Digital Sky Survey photographed 200 million celestial objects in the first five years of its operation. By 2020 it is hoped that 20 billion will have been photographed.

---

[6] Named for the American philanthropist and former CEO of General Motors Alfred P. Sloan, Jr (1875–1966).

Over 10 billion light-years (over 10^{26} metres)

The furthest object we can see is a quasar (quasi-stellar radio source) some 13 billion light-years away. Quasars are the oldest known bodies in the universe, and some of them are the brightest and most massive, outshining trillions of stars. A quasar is a halo of matter surrounding and being drawn into a black hole. So long as there is matter in the vicinity, a black hole will grow in size until it has cleared all material from the area of its gravitational influence. While it consumes matter in this active quasar stage it will shine brightly. Effectively, quasars are rotating black holes that are being fed with matter, which is why they are not only not black but very bright.

*

Here, then, is our universe: between 30 and 50 billion trillion (between 3×10^{22} and 5×10^{22}) stars arranged in 80 to 140 billion galaxies. These billions of galaxies are, in turn, arranged in clusters, clusters of clusters called superclusters, and as filaments of superclusters like the Great Wall. A precocious child might write her address as: the earth, the solar system, Orion Arm, the Milky Way, the Local Group, Virgo supercluster. If this is all the universe is – an oppressive collection of stars arranged in just a few structures – we might allow the universe its ridiculous size as a trade-in for our complexity. Yet can we say that we have got any further along? How are we to account for our presence among these structures of stars? And if these are the largest structures in the universe, what lies beyond them? We have arrived at what seems to be the edge of the universe with no clearer understanding of how the universe could have an edge. We cannot rest here.

Measure for Measure

Space and time seem to have a precarious existence in the minds of so-called primitive people and only harden with the idea of measurement.

Carl Jung, *Synchronicity*

Clearly the universe was not measured out by explorers setting forth with rulers. Mankind has barely stepped into outer space, in astronomical terms anyway. What we know of the universe has mostly come to us from out there. It is not we who have gone into the universe but the universe that has come to us, in the form of light.

We believe the universe is as we describe it because we believe in the means by which we measure and describe, and because we believe that the reality out there is consonant with reality as we have found it to be locally here on earth. We believe in the scientific method. But what is the scientific method and what are we actually doing when we make a measurement?

From the earliest times, mankind has tried to measure time and space. We see the world as made up of separate things that are connected temporally and spatially. The world is things that move. This is our starting point: not something we have to find out about the world, but something we believe

43

to be irrefutably how the world actually is. Some Eastern ways of thinking tell us that the opposite is true: that there are no things at all, only an indivisible unity of integrated phenomena, but such an impression of reality is hard won and almost as rare as Buddhas. It does not seem to be our natural response to the world. What we feel certain of – as certain as we are of the existence of our own self (another illusion, say the philosophers and mystics) – is that space extends (and there are separate things in it), and that time flows (allowing things to reappear in different parts of space). For most of us, who labour to exist without the leisure to contemplate that maybe we do not, the external world all too ponderously weighs down on us framed in space and time. The self may be a figment of the imagination, as the eighteenth-century Scottish philosopher David Hume (1711–1776) asserted, and time and space an illusion, as his near contemporary the German philosopher Immanuel Kant (1721–1804) argued, but to live in the material world is to live as Dr Johnson[1] did: when we see a rock we know we can kick it.

Overwhelmed by time and space we may be, but defining what we mean by time or space turns out to be more problematic.

We can imagine, since it is a notion we still struggle to free ourselves from, that mankind, when it first set out to measure the world, might have taken itself to be at the centre of things. We could blame the ego, or we could put it down to the fact that we look out at the world from here, where 'I am', which explains, perhaps, why egocentricity seems to be our natural condition. It is hardly surprising then, that kilometres, metres and centimetres, or miles, yards and inches, or any of the earliest units of measurement, are peculiarly suited to life on earth, since they were chosen to relate to the human body and

---

[1] The Irish philosopher Bishop Berkeley (1685–1753) wondered if an unobserved tree could be said to exist. In response, the English wit, critic, essayist and lexicographer, Samuel Johnson (1709–1784) said 'I refute him thus' and kicked a rock.

the activities of human bodies in the world, when the world meant the earth. The word 'foot' gives the game away. The origin of the yard is not known, though a popular account has it as the distance between the nose and thumb (of the outstretched arm) of the English king Henry I (1068–1135). But it could just as well have been, as is also put forward, the measure of a waist or a stride, or twice an ancient measure called a cubit. The Egyptian hieroglyph for a cubit is depicted as a human forearm, which is how this measurement of length was defined. An ell, once used by tailors, is said to have started out as the length of the arm measured from the shoulder to the wrist. The English, Scottish, Flemish and Polish ell are all different lengths.

Most units of measurement were first made to suit human beings, which is why we can tell when the temperature changes a couple of degrees, feel the difference of a few pounds per square inch of air pressure on the ear drum, comfortably hold a pound or two of weight in a single hand, and so on. But from the moment mankind began to measure, measurement was a problem in need of a solution. How are we to agree that when we make a measurement we make the same measurement? Today the problem hardly seems apparent. When we measure length we know what a metre is, even if we might be hard pressed to define it. Indeed defining it turns out to be an intractable problem, closely related to our inability to define what time and space are.

Ancient civilisations did not, of course, use the metre or the yard, but the problem remains the same. A block of black marble discovered in Egypt that is a cubit long and dating from 2500 BC appears to be evidence of an early standard measure, the earliest known unit used to measure length. Such a block of stone existed to ensure that there was local agreement on how the measurement of length was to be made, and was presumably referred to as a final authority to ensure that all cubits meant the same thing. To much the same end, King Edward I (1239–1307) ensured that all English towns possessed an official measure called an ellwand, also known as the Belt of Orion. Agreeing on a measure allows us to move from 'I am here. I am at the centre. I am the authority in all things', to 'We are here. We are at the centre of all things. We are the authority in all things', which is, at least, a step away from egocentricity.

Mankind did not come to any global agreements about measurement for thousands of years. There have been countless different measures across many different cultures and nationalities for countless different kinds of substance from gold to apples. England made no attempt to unify its different measures until the thirteenth century. Until 1824 there were still three different gallons, for measuring ale, wine and corn. The inch was measured differently in America and England until as recently as July 1959 when 2.54 centimetres was agreed on, though neither country went so far as to adopt the metric system.

Scientists, at least, have agreed that when they measure distance they measure it in metres. Occasionally someone forgets, as happened in 1998 when the Mars Climate Explorer crashed into the surface of the planet because an outside source supplied NASA with a measurement of the craft's position in miles rather than in kilometres, a mistake that cost $125 million.

The first attempt to define a metre was made in France in 1793, when it was defined as the ten-millionth part of that segment of the earth's circumference that runs from the equator to the North Pole, via Paris.[2] Even the least scientifically minded reader might find this definition suspiciously, well, French.

Ultimately, science searches for descriptions that can be agreed on across the universe not just around the world. Science is based on the belief that no matter where we might be in the universe, the reality we perceive, whatever we think it is, is the same reality. The ancients did not make this assumption: reality was, for them, divided into different spheres of

---

[2] The earth is not a perfect sphere so the distance would be slightly different if some other direct route was chosen that did not go via Paris.

influence. The world that included the earth and extended as far as the moon (the sublunar world) had a quite different reality from that of the heavens beyond: different laws of nature applied. Modern science works on the belief that there is one indivisible reality that is universally consistent. It is imperative that when we, as scientists, attempt to describe this reality by measuring the things in it that we all agree what we are measuring with.

This 'we' who describe the universe is a strange inclusive group. We earthlings have not yet travelled far from home, nor do we know whether or not we are the only beings in the universe who have embarked on such a description of nature, but science believes that there is a universal perspective. It either imagines that humans will one day be far-flung across the universe, or that there are other life-forms already there – aliens capable of describing the world as they apprehend it outside of themselves – who have embarked on the same scientific enterprise as earthlings have. No wonder scientists are peculiarly interested in aliens and in science fiction. The idea of aliens is almost as important as their actual existence. Scientists need an alien perspective in order to ensure human bias is eliminated. Yet what that perspective might be and what form other life can take is limited by our human ability to conceive them.

If there are aliens out there measuring reality with a stick, then we need to convince ourselves that the way we define length is universal. If we don't have such universal agreement, there will always be the possibility that when an alien describes reality something quite different from our reality will be described. And then who is to say which is the true reality?

The 1793 definition of a metre is not even global – there is privilege attached to being Parisian – let alone universal. Even if we could persuade all living forms in the universe to accept that a particular fraction of the earth's circumference passing through Paris is how we define our unit of measurement, we could only do so by asserting our authority, the ultimate assertion presumably made by war.

This early French definition failed, anyway, for a more mundane reason. It did not take account of the flattening of the earth due to the earth's rotation. The first prototype of the metre, made in 1874, was short by 0.22 millimetres. This oversight points to a deeper problem: even if we had, at the time, taken account of the earth's flattening, that flattening changes slowly over time. If, somehow, we had managed to persuade all beings in the universe to adopt the Parisian definition of a metre, it still wouldn't be universal through time, even if it could be made to be across space.

A new prototype was made in 1889, and the metre was given a new definition in 1927, and again in 1960 when it was defined as 1,650,763.74 wavelengths of the orange-red line of the spectrum of krypton-86 as measured in a vacuum. Such a definition may be very precise, but it is about as arbitrary and cumbersome as a definition can be. If we really believe the universe to be a place ultimately fashioned by elegant laws, which is indeed the curious faith we have in the mathematics that underpins all our scientific descriptions of the world, then we are unlikely to be satisfied with a definition of something as important as the unit that measures all of space, and the size of everything in it, that is so ugly. Since 1983, the metre has been defined as the distance that light travels in a vacuum in 1/299,792,495 of a second, which may seem hardly more convincing. And yet what this latest definition has going for it is that, finally, here is a definition that can claim to be universal.

For the moment, we believe that the speed of light is the same wherever it is measured in the universe. Putting our faith in this constancy, we feel sure that when we use a measuring stick defined by the speed of light we can all agree (across the universe, who or whatever we are) that any measurement we make will be the same measurement.

Alien life-forms, we can allow, are unlikely to have chosen the metre as their unit of measurement. If, however, we assume that they have evolved sufficiently to discover that the speed of light is a universal constant, then in theory we can agree –

through some simple translation of our unit into their unit – what reality looks like when we take our rulers to it.

There are problems even with this definition. In recent years, some doubt has been cast on the idea that the speed of light is constant, which means that this definition, too, may be culturally transitory. This leaves open the possibility that reality may still look quite different to alien life forms, either because they are more evolved than we are, or have come at reality in a different way.

The complications don't stop there. Our best definition of a metre comes out of our most sophisticated scientific discoveries and descriptions, which in turn have been derived using a definition of a metre which was not, and perhaps still is not, universal. Within the definition of the metre we find embedded all of science and the history of science. We find ourselves trapped in what appears to be circularity. Whether this circularity is real or apparent is a matter of philosophical argument (and so as far as many scientists are concerned, a pointless argument). Practical scientists might argue that scientific progress *is* the progressive refinement of measurement; philosophers might argue that science defines progress in its own terms, which is no way out.

Scientists measure the universe using rulers and clocks. Our current definition of length is dependent on knowing how to measure time: a metre is how far light travels in a minuscule fraction of a *second*. So if we want to know what a metre is we had also better have a good idea what a second is. But what we think we mean by time is even more difficult to pin down than what we think we mean by space. Time flows, but what flows? One moment becomes another moment, but how? And what is a moment? Why does time only appear to flow in one direction: into the future? Is time even linear? Its circular nature is in some ways more apparent.

The Greek philosopher Heraclitus (*c.*535–475 BC) had a go at defining time. One of the few fragments of his writings to survive reads: 'On those stepping into rivers staying the same other and other waters flow.' This statement is thought to refer to the flow of time or the flow of existence, and is more usually freely translated as: 'No man steps into the same river twice, for it is not the same river and he is not the same man.' Either way, Heraclitus tells us that, though the waters of a river continually change, the river also stays the same, an idea not unrelated to the Heraclitan fragment: 'Change alone is unchanging.' Parmenides, a Presocratic[3] philosopher alive in the early fifth century BC, thought that time is an illusion and that the deeper reality is eternal and unchanging. Most Greek philosophers were of the opinion that time was not created from anything: it just is. The philosopher and theologian St Augustine (AD 354–430) believed time to be a subjective experience: 'If nobody asks me then I know; but if I were to desire to explain it to one that should ask me, plainly I know not.' Leibniz took the view that time and space do not have a fundamental existence but are merely the means to describe the relationship between things. Immanuel Kant described time as a quality of the mind that orders our perceptions of the world. He was sceptical of the idea that there is a world that exists in time and space, and of selves that experience such a world. In the twentieth century the American physicist John Wheeler (1911–2008) defined time as that which 'keeps everything from happening at once'. Whatever the philosophical considerations, none of this helps the scientist in need of a pragmatic definition, rather than a description, of time.

Presumably the need to measure time, whatever it might be, was first felt about 12,000 years ago when mankind began to farm. Planting and harvesting are best done at certain times of the year. To carry out these activities with the greatest efficiency it would be a good idea to know in advance when these periods are to occur, which is what a calendar allows.

The first calendars were based on astronomical activity. A day measures the time the earth takes to spin once on its axis. A year is the time the earth takes to make one complete orbit of

---

[3] Those philosophers, like Heraclitus, who came before Socrates (c.470–389 BC).

the sun. There is no reason to suppose that days and years fit together in any simple way, as indeed they do not. The history of the calendar is one of making them fit, and also to fit together with the complicating motion of the moon. Its twice-daily creation of the tides and its monthly effects on biological rhythms make the moon hard to ignore. Though these phenomena are celestial – in contrast to our early earthbound attempts to define length – the moon and sun are, an alien would tell us, local phenomena. As scientists, we are determined to find a perspective that includes earthlings and aliens. Scientists believe in a world that can be seen as it actually is: separate from ourselves, out there, the same world that can be experienced in the same way no matter where or what we are in it.

The history of the calendar is as determined by culture as those early definitions of length. The Julian calendar went unmodified from the time it was introduced by Julius Caesar in 46 BC, to reform the Roman calendar, until it was regularised under the name of Pope Gregory on 4 October 1582, a date that was immediately followed by 15 October. The adjustment was required in order to account for the drifting over the centuries of the timing of Easter. England and America resisted the Gregorian change until Wednesday, 2 September 1752, and Russia, until 1917. The Gregorian calendar is accurate to 26 seconds a year, or a day every 3,323 years.

Scientists treat time as if it is a dimension like space that can be cut up into little pieces. Science requires time to flow in regular measures. Without it, Isaac Newton (1643–1727) would not have been able to formulate his laws of motion. An accurate calendar suggests that time has such a nature: that it can be measured in lumps like months, days, hours or minutes. Time was not always treated in this way. In Europe, until the fourteenth century, it was common to measure the day as the time the sun rose to the time it set, with the day divided into hours accordingly. The result was that day-time and night-time hours were different from each other, and progressively different as the year moved forward. Night hours would have felt very long in the winter because they *were* very long.

There were clocks in China from the eighth century, and mechanical clocks in Europe from the beginning of the fourteenth. The first mechanical clocks were huge devices housed in church towers. These clocks make use of an escapement, a device that gradually and smoothly translates the wound-up rotational energy of a spring into the oscillating motion of a balance-wheel or pendulum. The first pendulum clock was invented and patented in 1656 by the Dutch scientist Christiaan Huygens (1629–1695). It was Galileo (1564–1642) who, in 1602, had first studied the motion of a pendulum. He realised that the uniform ticks of a pendulum's motion could be used for keeping time. From 1637 he began to explore the idea of making a pendulum clock, but he died before he could turn the idea into a physical reality. The physical movement of a pendulum allows us to believe (whether or not it is actually true) that time does after all exist as something that flows evenly and can be divided up, in the same way that we believe space can be divided up and measured off in regular amounts. Linear time may be the single most significant reason why the scientific revolution happened in the West and not in the East (though during the Renaissance there was a widespread debate about whether time was linear or circular). In broad terms, in the East, and also in so-called primitive cultures, history was taken to be an endless cycle of repetition. Even today, the Hopi people and some other Native American tribes speak a language that avoids all linear constructions in time. First in the Western world, and now throughout the world, time unfolds in a line that stretches into the future, and along which we mark out what we call history and progress.

Pendulum time is different from celestial time, as is suggested by the cyclical nature of the seasons. Specifically, in scientific terms, it is clear that celestial time is not linear. The earth is furthest from the sun on 4 July and closest to it on 3 January. The velocity of the earth speeds up as it gets closer to the sun, so the time from sunrise to sunrise changes. There can be no regular tick in celestial time as there is in pendulum time. Pendulum time allows us to envisage an artificial unit of time like the second, which is quite different from real (if local) measures of time like the day and year.

Nevertheless, the first definitions of the second were attempts to relate it to real time. A second was once defined as 1/86,400 of a solar day, but this definition is not universal, and the solar day has not remained constant. The earth used to spin more quickly on its axis and what we call a day would have been several hours shorter half a billion years ago. This definition of a second proves to be an historical choice and a cultural one.

The second has had several other definitions, but in 1967 it was settled that it should be defined as 9,192,631,770 periods of radiation of a very fine transition between two energy levels of the caesium-133 atom. Clearly, this definition is as arbitrary as our definition of length.

When we make measurements, we don't seem to be able to separate out the unit of measurement from our own nature. But such philosophical conundrums do not trouble most scientists. ('Shut up and calculate!') Though we cannot be sure what we are doing when we measure time and space, science goes ahead and measures them anyway. The history of science is the history of more precise measurement, which in turn allows the units of measurement to be more precisely defined. The scientific method is more than a philosophical dilemma: it works. By 'works', we mean that there is a technological reality that is our evidence – what we call progress – and which we live in and call home. If the idea of progress is an illusion played out on an imagined line of time, it is certainly a compelling illusion, the name of which is materialism.

Throughout the history of science, the universe scientists describe has grown in size and age. Though it appears to reduce our units of measurement to insignificance, we cannot be sure that the universe does not take a second to be a very long time, or a metre to be a very long length. There is a danger that if the units themselves are seen as homey, then our description of the universe is askew because of some biased idea we have about what home is. Scientists are trying to find universal descriptions of the world – indeed universal descriptions of the universe – not homey ones that are culturally or historically biased. They try to take measurement away from human beings to something invariant. Whether this is actually possible is a question that, perhaps surprisingly, remains open. What convinces is that out of scientific measurement (ultimately reducible to the measurements of clocks and rulers) we can construct rich theories that describe multiplicities of phenomena that characterise the world out there. And so we become less interested in the shifting foundations on which that knowledge is built. We believe in electricity not because we know what it is (not really, not deep down), but because the scientific description of it explains a great deal of phenomena. Electricity fits into an overall story that is consistent with our belief in other phenomena – magnetism, say. We can build up a web of description, the strands of which are stronger for being a part of the whole web, that captures and holds more and more of what we call material reality.

Systematic measurement is what science does. Science measures the universe and its contents, and by measurement we mean any act of observation of what we take to be the outside world. Science formalises observation as experiment that, crucially, can be repeated. A scientific experiment isolates an aspect of reality, observes it, and makes it public. In principle any experiment is repeatable. In practice it can be very hard. For the moment, trust and the peer-review system keep the integrity of the scientific method intact.

In order for something to qualify as an object of scientific enquiry it must be possible to reconjure that object and remeasure it. An experiment is designed to isolate an object, to separate it from the rest of the universe, so that it can be measured. It is the act of separation that makes something the 'it' that can be measured.

The scientific method is about division, about dividing the world up into parts and giving those parts names, as a first step towards describing how those parts interact with each other. The word science is derived from both the Old English word *sceans*, meaning to separate out, and the Latin word *sciens*, meaning to know. But to believe that science itself is divisive would be to confuse the methodology with what the methodology uncovers. Science separates out in order to seek out better descriptions of a unified reality. The parts fit together.

You believe there is something because you know you exist. You believe in your own ego. Science is a way of translating that individual experience of the world into a collective experience. We can personally validate a scientific description of reality by repeating an experiment, or by believing that experiments are repeatable, or, most apparently, by simply noting the changing nature of the world that technology creates around us. Technology is our evidence that science is getting somewhere. And by somewhere we mean our ability to create simulacra of reality that are the material world. The steam engine, drugs, central heating, weapons, particle-accelerators and i-Phones all convince us that the world is real, and becoming somehow more real, the more sophisticated that material reality becomes. Sometimes we forget that, no matter how elaborate the material world has become, nature must be more elaborate since the material world is a sieved-out part of it. A hard-line materialist might claim, as a matter of faith, that science will ultimately pass all phenomena through the sieve of the scientific method.

Some phenomena are hard to reproduce. It could be argued that most phenomena are extremely hard to reproduce. Scientists rule out certain phenomena as unworthy of, or unsuitable for, scientific investigation: phenomena that can't be isolated, that are not publicly reproducible. What, for example, are we to make of the alterations of love? Do such phenomena occupy some realm separate from the realm of science, to be apprehended only by poets and mystics, or are they awaiting a material description? The novelist Hilary Mantel remarks:[4]

---

[4] In the Introduction to *Who Is It That Can Tell Me Who I Am?* by Jane Haynes (2007).

Our whole world of experience comes to us subjectively, but that doesn't mean we can't make valid statements about it. We just need to differentiate between qualities that can be measured and qualities which can't – without stigmatizing the latter as less useful. The heart's electrical pattern can be traced, but not the wayward impulses of love and hate. Yet who could maintain that the latter don't have real effects in the world? Those who do not believe in what cannot be measured or quantified are on shaky ground: their inner reality is doomed to be alarmingly divorced from the reality of most of those about them.

Curiously, the one fact that we think we know of the world – the certainty of our own existence – is not open to scientific examination, as it is by definition not public. If the goal of science is to search out a description of everything, then, ultimately, all forms of knowledge must somehow merge. The mystery of a material description must become indistinguishable from the mysticism of, say, a poetical description. Or perhaps there are two worlds fated to remain separate: 'Our feelings belong to one world, our ability to name things and our thoughts to another; we can establish a concordance between the two, but not bridge the gap.'[5]

Technology progresses because the theories that describe the world we measure become more sophisticated, that is, they encompass ever more phenomena in a single description, and theory progresses because we find out ever more sophisticated mathematics in which to write these theories. Why nature is describable in mathematics is perhaps the greatest of all science's mysteries. The buck stops there: our ultimate scientific faith rests in mathematics and the web of phenomena that mathematical description includes. Technology is the outward and visible sign of that faith. We no longer believe in human perfectibility, or the lessons of history, or any other form of

---

[5] Marcel Proust, *In Search of Lost Time* (1909–1922).

progress, but scientific progress remains because the technological aspects of our lives change. The comfort of the material world has allowed man to retreat indoors to become
physically and philosophically removed from nature. Progress
is a feedback loop between technology, theory and mathematics. Newer and deeper theory is written in newer and more
refined mathematics. Armed with the latest technology, a scientist scours the world looking for evidence to back the theory
up. Out of a greater understanding of the material world comes
the ability to make ever more refined measuring instruments.
Through technological devices such as the telescope or microscope, our ability to sense the world is extended. Or rather
our sense of sight is extended, since on the whole we do not
smell, taste, feel or even (despite the so-called Big Bang) hear
the universe.

Science is consistent measurement. We expect the world to
look the same when we measure it again, or, indeed, if anyone
else makes the same measurement. Science requires repeat-
ability. Phenomena that arise out of rogue conditions through
the mediation of rogue individuals are not suitable subjects
for scientific enquiry. But we are all rogue individuals. Complex,
individualistic humans are always going to be the most
intractable objects of scientific enquiry. It is our own nature
that presents the greatest measurement problem.

The universe itself would seem to be outside the remit of
science, since what is there to measure the universe against
except itself? There are no other universes with which to
compare our universe. In practice, the universe is always in the
process of being redefined. There is always a newer, enlarged
idea of the universe to which the older, 'smaller' universe can
be contrasted. In the computer age we can simulate other
possible universes as computer models. It has even been
suggested that we might some day create other universes like
ours, in a process that must necessarily downgrade what we
have called the universe, to something local. Ironically, if science
ever achieved its aim of describing the oneness of nature,
whatever that oneness was it could not be a scientific object.

A single unified description of the universe would necessarily bring the kind of science that compares one thing to another to a full stop.

But such a description seems to be always just beyond our reach. The universe is a chimerical object that changes into something entirely *other*, the closer we think we are to grasping the whole. The universe would appear to be forever more creative than our most creative abilities to think about it, which is hardly surprising if we see ourselves as part of the universe's own output, 'a hope-filled and hopeless striving of life to comprehend itself, as if nature were rummaging to find itself in itself – ultimately to no avail, since nature cannot be reduced to comprehension, nor in the end can life listen to itself'.[6]

---

[6] Thomas Mann, *The Magic Mountain* (1924).

It's Not About You

Not at first did the gods reveal all things to mortals, but
in time, by inquiry, they made better discoveries.

Xenophanes

Our understanding of how the contents of the large-scale
universe are arranged – as a hierarchy of stars in motion – is
the result of hundreds of years of scientific investigation.
Whatever the scientific method has become, it was not always
as it is now. It has evolved over time, in tandem with our
understanding of the universe, and doubtless will continue to
evolve as our understanding of the universe deepens. Science
and the universe are inseparable.

To get to the edge of the universe we must take a long route
through history. To answer those nagging questions: where did
the universe come from? and what is made out of? we need
to go back to the beginning of the scientific venture in order
to find out how we have arrived at our current understanding.

In broad terms, both science and the universe have an ancient
and a modern past.

Science is a collective endeavour, without a written consti-
tution, whose meaning has emerged over time. Whatever
science is now, its history stretches back to a time when the

word was meaningless. Today we know how the stars are arranged, which makes it easy enough to see that we are not at the centre of the universe. But it wasn't always so. Ancient science started out with the opposite idea. By the time of Aristotle (*c.* 384–322 BC) the earth was firmly fixed at the physical centre of the universe, part of a cosmological description that has a history stretching back to the beginnings of whatever we mean by civilisation.

Whatever we mean by civilisation appears to have arisen in city-states in the Near East. The most influential ancient civilisation (in Western history) was in Mesopotamia, a fertile region bounded by the Tigris and Euphrates rivers in what is modern-day Iraq. The word Mesopotamia is derived from Greek words meaning between two rivers. There is evidence that there was farming here from 10,000 BC, around the time the earth became as warm as it is today, and warmer than it had been for almost 2 million years. Nomadic humans, who had moved around in groups of between 20 and 30, settled down as communities that began to grow in size. There is evidence that by 7000 BC there was a fortified farming community at Jericho of some 74 acres.

A tribe called the Sumerians arrived in Mesopotamia around 5000 or 4000 BC, but it is not known from where. The Sumerian society was the first to learn to read and write. The oldest known story, dating from the third millennium BC, is the *Epic of Gilgamesh*, a summary of legends from Babylonia, a state in the south of Mesopotamia. It is the story of the King of Uruk and mentions many of the first city-states around which civilisation evolved: Ur, Eridu, Lagash and Nippur. It also contains the first account of a great flood, and the first account of a dream. In the Bible we are told that Abraham, the father of the Hebrew and Arab nations (the Israelites were descended from his son Isaac, and the Ishmaelites from his son Ishmael) travelled from Ur of the Chaldees. Chaldea was a region of Babylonia.

A creation story from perhaps the eighteenth century BC, called *Enuma Elish*, tells of the creation of Mesopotamia and of man. It was recited in temples for hundreds of years. These first creation stories are both the first religious accounts and the first cosmologies. Aspects of the *Enuma Elish* were absorbed into Hebrew cosmology and the biblical account of creation. The earth is a flat disc surrounded by water, above and below. The firmament keeps the water above from deluging the land, but allows rain through. From below, the water rises up as rivers and seas. The Sumerians studied the skies as astrologers and astronomers. They could see portents from the gods and predict eclipses.

Other civilisations were also developing across the world: in Egypt from around 3000 BC, the Indus Valley from 2700 BC, China from 2100 BC. But for whatever reasons (and many reasons have been put forward) the history of science is largely a story that came to be told in the Western world. Eastern modes of thought seem to be in opposition to the idea of progress that is at the heart of science. It has been suggested that the Chinese pictogram did not encourage abstract thinking, with the result that, as the philosopher John Gray tells us: 'Chinese thinkers have rarely mistaken ideas for facts.'[1] The Egyptians and Babylonians seem to have left nothing that could be counted as a description of the material world, though the Babylonians did develop a counting system based on the number 60, a legacy we see today in the 60 minutes that make an hour and the 360 degrees that complete a circle. The Egyptians had a calendar that was based on observation of the stars. The earliest recorded date in history we know of is Egyptian: either 4236 or 4241 BC according to how their calendar is interpreted.

There were Greek tribes adrift in the Aegean from 2000 BC, eventually settling as city people. The so-called Mycenaean civilization flourished from around 1600 BC before collapsing in 1150 BC. Greek history entered a Dark Age for some 300 years. The Olympic games were founded in 776 BC, and Homer

---

[1] John Gray, *Straw Dogs* (2002).

(who may have been a tradition rather than a single writer) cannot have been around before the eighth century BC. And so began the civilisation the German philosopher Friedrich Nietzsche (1844–1900) called 'the most accomplished, most beautiful, most universally envied of mankind'.[2] Some historians say that Greek philosophy started on 28 May 585 BC. It was on this date that Thales of Miletus (*c.*624–*c.*546 BC), the first of the Presocratic philosophers, is said to have predicted an eclipse. It is now thought that Thales observed the eclipse rather than predicted it, and that his knowledge of it was passed down from the Babylonians. Chaldean wise men travelled abroad taking their knowledge of astrology and early astronomical observations into the Greek and Roman empires.

Whatever claims were made for him, Thales certainly eclipsed his precursors, and for this reason he is often called the father of science. It was he who introduced the word *cosmos* to describe the universe, which as well as being the Greek word meaning order also has the meaning of something that beautifies, as in the word cosmetic, and is the opposite of the Greek word *chaos*.

Thales believed that everything is made of water in one form or another. It was he who began the search to find the physical components that make up the world: materialism started here. He made no distinction between the living and inanimate. For Thales, even magnetic rocks possessed soul, a mystical idea that persisted into the sixteenth century in the work of the English physician and scientist William Gilbert (1544–1603), who was an ardent and early supporter of the sun-centred Copernican model of the universe, and one of the first to use the word 'electricity'.

Early Greek philosophy can be traced through teacher and pupil in a mentoring tradition. The word 'mentor' is taken from Homer's *Odyssey*. Mentor stands in as father to Telemachus when his real father, Odysseus, is away at war. Thales was

---

[2] Friedrich Nietzsche, *The Birth of Tragedy* (1872).

mentor to Anaximander (*c.*610–*c.*546 BC), and Anaximander to Anaximenes (*c.*585–*c.*525 BC), all three of whom came from Miletus, an ancient town in what is now modern-day Turkey, then part of the Greek world. Anaximenes continued the search for simple descriptions of the world. Rather than water, he said that air was the principle from which everything sprang.

The most famous of the Presocratics is Pythagoras (*fl.* sixth century BC), the first to call himself a philosopher, literally a lover of wisdom. He had studied with wise men in Egypt, and later in Phoenicia (an ancient civilisation that is now the coast of Lebanon and Syria). It may have been in Egypt that he became interested in geometry and trigonometry.

Pythagoras founded a school that lasted a millennium, though the Pythagoreans were perhaps more a brotherhood than a school. Named *mathematikoi* (literally, those who study everything), they were strict vegetarians, living a monk-like existence. They made a particular study of arithmetic, geometry, music and astronomy: the basis of education until well into the Middle Ages, by which time it was called the quadrivium[3] (a Latin word meaning where four roads meet). They believed that at bottom reality is mathematical, a belief that has persisted to the present day. An important distinction is that though we take shape and number to be attributes of things, for the *mathematikoi* shape and number are the essence of things. Numerology was part of the Pythagorean tradition. In the ancient world the modern distinction the scientific method makes between mysticism and mystery was meaningless. Numerology is at the heart of the book of Chinese divination the *I Ching* (probably compiled in the ninth century BC but mythologically dated to 2800 BC), and the body of Jewish esotericism called the Kabbalah dating from around AD 1000.

For the Pythagoreans, the most perfect shape of nature was the circle. Pythagoras put the earth at the centre of a spherical

---

[3] The trivium (where three roads meet, and from which we also derive the word trivial) was grammar, rhetoric and logic. The quadrivium and trivium together make up the seven liberal arts.

universe, and simple numbers were used to describe the motion of some of the known planets. Because he did not leave any writings (nor are many writings left by his followers), what has been ascribed to Pythagoras has been much debated. We only know about Pythagoras from contradictory accounts written 200 years after his death. It is now known that he did not discover the theorem that is named after him,[4] nor the relationship between musical intervals and simple numbers that is usually attributed to him.[5]

Heraclitus (*c.*535–475 BC) described how the cosmos is created out of pre-existing chaos. The cosmos is order imposed on chaos, which we witness as the material world. The ordering principle is called *logos*, from which the suffix -ology is derived. *Logos* is sometimes translated as 'word', as it is at the beginning of the English translation of the original Greek version of the gospel according to St John: 'In the beginning was the word.' Chaos is a condition in which there are no things, a world in which whatever there is is without a name. No thing and nothing are quite different ideas. It is the naming that makes the separation out of chaos into cosmos. This was clearly the original meaning of the account of creation in Genesis, when God separated out from chaos what then became things with names (light, the earth, Heaven, night, day, and so on). Medieval theologians imposed on this creation story the idea that the world was created *ex nihilo* (out of nothing).

Heraclitus wrote that change (also characterised as fire) is the fundamental quality of the world, an idea that resonates with the modern understanding that everything is a form of evolved energy.

Fragments of a single poem are all that have survived of

---

[4] Pythagoras' theorem, as every schoolboy knows, tells us that in right-angled triangles the square on the hypotenuse is equal to the sum of the squares on each of the other two sides.

[5] For example, a string can be plucked at the octave by dividing it in half. A fifth is found by dividing the string in the proportions 3:2, and a fourth in the proportions 4:3.

the philosophy of Parmenides (*c.*510–*c.*450 BC). He wrote that existence is eternal and unchanging: what we perceive as change, as in the motion of things, is an illusion. He denied the existence of nothingness and wrote that reality is an unchanging whole. His ideas influenced the philosophy of Plato, who acknowledged him as 'our father Parmenides'. His philosophy was reduced to the Latin tag *ex nihilo nihil fit* (nothing comes from nothing).

Empedocles (*c.*490–430 BC) synthesised earlier philosophies. For him the cosmos is made of earth, air, fire and water, and two principles: attraction and repulsion, also seen as love and strife. These four elements were the building blocks of the material world until the time of the European Renaissance.

Leucippus lived in the first half of the fifth century BC. Nothing survives of his own writings, and we only know of him because he was the mentor of Democritus (*c.*460–*c.*370 BC), who propounded the philosophy of atomism, which he may have taken from his teacher. Aristotle was an admirer of Democritus, and it is only because Aristotle criticises Democritus' atomism that we know about it at all. Again, mere fragments of Democritus' vast output survive and his work is mostly known of through the writing of others. Atomism tells us that everything is fashioned out of small, indivisible and eternally existing particles called atoms. Some atoms might, for example, have hooks on them and others be round. The differences between atoms, their different textures and shapes, and how they attach to each other, explain why different substances have different qualities. The atoms from which different foods are made affect the tongue in various ways, which explains the subjective experience of taste. The taste is not the essential quality of food. The essential quality is its atomic nature. Even the soul has an atomic structure, made out of the finest atoms.

Democritus was the first to assert that there are other worlds in other parts of the universe, with other suns and other moons.

Philosophy as practised by the first Greeks was the belief that wisdom is the essence of the cosmos. The Presocratics inherited a 2,000-year-old tradition of wisdom poetry from the Sumerians. If the Presocratics only occasionally presented their work in the form of poetry, it often has the force of poetry.

Ecclesiastes, Proverbs, Job, the Song of Solomon, and other wisdom books of the Bible were written around this time. Confucius (551–479 BC) was a near contemporary of Pythagoras. The Buddha is said to have lived from around 563 to 483 BC, though modern scholarship suggests a later date is more likely, on either side of 400 BC. According to Chinese tradition, the philosopher Lao-tzu lived in the sixth century BC, though he has now been dated to the fourth century BC by historians. It is conceivable that the Persian poet and prophet Zoroaster was alive in this era, though his dates are highly contested. He may have lived, though it is highly unlikely, as early as 6,000 BC.

*

Socrates (*c.*470–399 BC) was named as the wisest of all the Greeks by the Delphic Oracle, and was mentor of perhaps the most famous philosopher of all. The English mathematician Alfred North Whitehead (1861–1947) famously said of Plato (*c.*428–*c.*347 BC) that all contributions made after him are simply footnotes to philosophy. Plato founded his Academy in a grove of trees belonging to a man called Academos, hence the word, and the phrase groves of academe. The Academy existed until AD 529, that is, for over 900 years. Oxford and Cambridge universities received their charters in 1231. Not until around AD 2180 will they have outlasted Plato's school.

For Plato, the material world decays and disappears, and so is temporary and illusory. The real world, he argued, is a world of so-called ideals, and is eternal. The material world is an imperfect representation of these ideals. Perfect geometrical shapes, for example, exist in this Platonic world. The motion of the heavens is circular, as in the Pythagorean philosophy, because a circle is the perfect, idealised shape. The bodies of the heavens are spheres for the same reason. The knowledge that planetary orbits are ellipses and not circles still has the power to shock even today, so instinctively do we respond to the idea that the motion of the heavens must be circular as the ancients believed.

Plato developed Pythagoras' spherical universe as a series of nested spheres rotating inside each other with the earth at the centre. There were seven celestial spheres carrying the known planets and the moon. God was just beyond the seventh heaven. For Plato nature is impure. Perfect forms are not to be found there. How things really are can only be reached through reason, or wisdom. For him the cosmos is a place of order and goodness, a philosophy also inherited from Pythagoras. The universe is musical and has soul. It is dynamic and living. Plato was the first to ask why there is a universe at all.

Plato insisted on a mathematical underpinning of nature that was of little interest to his pupil Aristotle (c.384–322 BC). Aristotle was more interested in how the celestial spheres moved inside each other than in their ideal nature. There are 54 spheres in his cosmology, including an outer sphere that carried the so-called fixed stars. Aristotle took up the four elements of Empedocles' philosophy, and added his own fifth element: subtle stuff called aether (or quintessence), out of which the heavenly spheres and bodies are constructed. By medieval times aether had hardened into crystal.

For Aristotle the world of change happened in a region that extended from the earth to the moon. Beyond this sublunary sphere was the ethereal world of eternal unchanging things. In the world below, heavy objects fall to the earth because they have more earth in them than lighter objects, and so find their way back to where they should naturally reside. Objects with more airy natures, like feathers, would tend to be attracted to a more airy environment. Aristotle's accounts of the world are more discursive than the modern scientific method allows. To make such an account a modern and rigorous scientific description, we would seek to quantify the amounts of earth, air, fire and water that objects are said to contain, and would search for a mathematical relationship that unites the phenomena and makes predictions.

Like many disciples, Aristotle reacted against his mentor. Aristotle believed that the world was best understood by observing it. 'Nothing is in the intellect that was not first in the senses' is the motto that the thirteenth-century theologian Thomas Aquinas invented to describe Aristotle's methodology. Nevertheless, Aristotle's observations did not amount to scientific investigations of nature in the modern sense. He looked at the world from a distance and drew conclusions about how it must be. He did not look at the world closely, which is what we do when we perform an experiment. Aristotle claimed, for example, that men and women have a different number of teeth, though it takes only a little investigation of nature to show up the error. Aristotle's belief in a physically existing world that could be observed in order that it might be understood is, however, a step towards the modern scientific method. His method differs in that it emphasises human perception of how the world appears to be over investigation of how it actually is. For Aristotle, it was clear that heavier objects fall faster than lighter ones. It would take another 2,000 years of investigation of the world to show that this is not the case.

In the fourth century BC, Aristotle's most famous pupil, Alexander the Great (356–323 BC), captured Mesopotamia. The region had served as the hub of the Akkadian, Babylonian and Assyrian empires, but its historical significance had begun to fade by this time. In 331 BC Alexander founded the city of Alexandria. From the beginning of the third century BC a library was built there called the temple of the muses (from which we derive the word museum). The first librarian was called Demetrius, another pupil of Aristotle's. The library grew to be the largest body of knowledge in the world at that time, comprising perhaps half a million manuscripts. The great mathematician Euclid was active at the library in around 300 BC.

One of its most famous librarians was Eratosthenes (c.276–c.194 BC), who made the first accurate measurement of the circumference of the earth. For some time, the Greeks had known that the earth must be a sphere given that it casts a curved shadow on the moon. Using a piece of information brought to him by a traveller to the library – that the sun at midday shone directly over a well near Aswan – Eratosthenes realised that he could calculate the circumference of the entire world. Using the known distance between Alexandria and Aswan, the angle of the shadow cast by a marker at midday at Alexandria and the fact that there was no shadow at Aswan, Eratosthenes was able to calculate how much the earth curved between these places. From that piece of information it is easy to calculate how large the whole circle must be of which the curve between Aswan and Alexandria is a segment. That circle is the circumference of the earth.

He measured the circumference as 250,000 stadia, though there has been historical disagreement as to exactly how long a stadium is. Modern archaeological research suggests that if Eratosthenes had used an Egyptian stadium as his measurement then he may have been within 1 per cent of the true measurement (which is a little over 40,000 kilometres). This measurement was the most astonishing (most likely a fluke) of a series of accurate measurements made by the Greeks, measurements not repeated until modern times.[6]

The library burnt down when the city was attacked by Julius Caesar in 48 BC, but was rebuilt. Most of the contents perished in the third century AD on the orders of Emperor Aurelian. And in AD 391 manuscripts that had been hidden away were found and destroyed as part of the campaign of the then bishop of Alexandria, Theophilus, to raze all pagan temples. The last librarian was a man named Theon, father to Hypatia, a Platonist, mathematician, astronomer and high priestess of Isis. Hypatia was murdered – flayed by oyster shells[7] – by

---

[6] Christopher Columbus (1451–1506) ignored measurements made by Eratosthenes and others, arguing that the earth must be much smaller. He might never have set out if he had been persuaded otherwise.

[7] The Greek word is *ostrakois*, which also means roofing tile. So perhaps broken tiles were used to flay her. (The Greeks had a system by which citizens could be expelled by the casting of votes. The votes were written on roofing tiles, hence the word 'ostracise'.)

a gang of Christian monks in AD 415 at the age of 45. In AD 642 the last few remaining manuscripts were said to have been used as fuel to heat the baths of Egypt's Arab conquerors. The story is almost certainly apocryphal, probably put about by later generations to discredit the Muslim conquerors. By the end of the eight century AD the library's 1,000-year history had faded away to nothing.

Though the library at Alexandria was not the only depository of ancient knowledge – there was a rival at Pergamum from 200 BC – by the time the library was in its final decline, much of what the ancient world had learned had either disappeared forever or was about to be lost to the West for centuries. In the late fourth and early fifth centuries St Augustine worked Plato's ideas into a Christian belief system. The sixth-century Roman philosopher Anicius Manilus Severinus Boethius (c.480–524) devoted his life to the preservation of ancient classical knowledge, translating many Greek texts into Latin. He was one of the last scholars proficient in Greek before the West lost historical contact with the classical world. Boethius is sometimes described as the last of the classical writers. His masterwork, *Consolatio Philosophiae*, was written in prison while he awaited execution. It was translated from Latin into English (as *The Consolation of Philosophy*) in the fourteenth century by Geoffrey Chaucer (c.1343–c.1400), at a time when the Western world, particularly in Italy, began to re-establish its connection with the classical world.

The Renaissance – that great flowering of the intellect that followed the Dark Ages – marked not only the West's rediscovery of classical knowledge but in addition the discovery and synthesis of the body of knowledge that had grown up in the Arab world over hundreds of years. Baghdad had become the centre of the civilised world within a century of the death of the Prophet Muhammad (c.570–632), and that world was largely impregnable to the West. For centuries, much of what survived from the classical world was protected and added to in the Arab world. The story of science has largely been told as a story of the Western world, and 400 years or more of

Arab thinking has been sidelined. Sometimes the 'we' that science means to be universal isn't even global.

For a time, knowledge was Arab knowledge. It found particular expression as alchemy, from an Arab word *al-kimiya*, itself derived from an Egyptian word *keme*, meaning black earth, after the fertile black silt that is carried by the annual Nile floods. Alchemy studies the workings of spirit and matter as part of a unified system. It is only in modern times that the two systems have been separated. Newton wrote a million words on alchemy, including a commentary on the *Emerald Tablet*, a text that purports to reveal the secret of the transmutation of the cosmos's primordial substance into other forms. The *Emerald Tablet* was supposedly written by the Egyptian god Thoth (in his incarnate form of Hermes Trismegistus), and was once housed at the library in Alexandria. It was influential in the West and led to the development of a system of enquiry based on secrecy and obscurity called the hermetic tradition. Newton's contemporary Robert Boyle, the father of modern chemistry, was also interested in alchemy and hermeticism. His *Dialogue on the Transmutation of Metals* was lost but later pieced together from fragments. If nothing else, the etymology of chemistry can be traced back to alchemy.

During the Renaissance, many of the classical works that had been protected and interpolated by the Arab world were translated into Latin, not from the original Greek but from Arabic. For a period the art of translation was one of the high arts of the Renaissance. A collection of hermetical writings called the *Corpus Hermeticum*, Greek texts from the second and third centuries, was translated into Latin in AD 1460 by the Florentine philosopher Marsilo Ficino (1433–1499), who put aside his translation of Plato's dialogues in order to work on them. Florence was the centre of the humanistic tradition and of the Renaissance throughout the fifteenth century, and the *Corpus Hermeticum* was enormously influential for hundreds of years during and after the Renaissance. The philosophy of humanism – the idea that mankind is responsible for its own destiny – can be traced back to this body of work. Surprisingly, perhaps, humanism was not condemned by the Church. Rather the opposite: Christian and hermetic knowledge were synthesised as humanistic Christianity. Ancient Greek anatomies of love (eros, agape, pothos and himeros: the Greeks had words for it) were re-examined and integrated into a humanistic philosophy. Plato tells us of the rare regard in which Socrates held his pupil Alcibiades, a form of love that came to be known as platonic love, and which was re-expressed during the Renaissance as the love between man and God. Humanism does not deny God so much as assert the belief that, when it comes to the workings of the world, belief is not enough, what is required is rational thinking and observation. The laws of nature are God's laws, or they stand on their own. Either way, man might come to understand them by thought and measurement. The divine mind, on the other hand, is sought out and understood through contemplation.

For hundreds of years, the Greek language itself had been lost to the West. The Italian poet Petrach (1304–1374) had tried to learn Greek but failed. Dante knew of Homer but couldn't read him. The Italian writer Boccaccio (1313–1375) was one of the first to learn Greek in modern times, and he ensured that Greek was taught at the University of Florence. Greek was re-established in Italy by the mid-fifteenth century. In the first part of the sixteenth century, it was his study of Greek religious manuscripts that led Martin Luther (1483–1546) to reformulate Christianity as Protestantism.

In the thirteenth century the philosopher and theologian Thomas Aquinas (c.1225–1274) had almost single-handedly created a synthesis of Christian theology and Aristotelian philosophy. In the fifteenth century the Western world was dominated by the Catholic church and still firmly in the grip of Aquinean thinking. Aquinas's philosophical system survived well into the sixteenth and seventeenth centuries; indeed it could be said to have survived into the present day. Aristotle's cosmology was how the universe was described, with some modifications along the way, even at the height of the Renaissance.

The Church was the final authority in all things spiritual and material, and if the authority of God was first embodied in the Pope, its second embodiment was in Aristotle. To look it up in Aristotle was the unthinking end of most debate. Where Aristotle proved to be of no use to the Church was in accounting for the fact that Easter was drifting in the Church's calendar and no one seemed to be able to do anything about it. After 1500 years, the vernal equinox had moved from 21 March to 11 March. (Solving the calendar problem is part of the history of science, but the search for a solution arose out of the history of Christianity.)

There were hopes that the rediscovery of Ptolemy's lost works might help solve the problem. Aristotle's cosmology had been enlarged on and somewhat improved by Claudius Ptolemy (c.AD 100–170), an Egyptian astronomer working in Alexandria and writing in Greek. A ninth-century translation into Arabic of his major work, the *Almagest*, had only mythological status in the West; a twelfth-century Spanish translation and a later Latin translation both failed to render many of the technical aspects of Ptolemy's cosmology. It wasn't until the rediscovery of the Greek language in the fifteenth century that Ptolemy's work began to make an impact.

Ptolemy had been both an astronomer and a mystic. Like Aristotle, he placed the earth, and thus mankind, at the centre of his cosmology, and it is likely that Ptolemy meant to place mankind at the spiritual centre of the cosmos too. He also seemed to be aware, in a modern way, of mankind's insignificance in the face of an overwhelming universe. He wrote that the earth, though centrally placed, could be taken to be nothing more than a mathematical point (that is, without size or dimension) in relation to the universe as a whole.

It isn't known how original Ptolemy's ideas were. He appears to be greatly indebted to Hipparchus (190–120 BC), who lived three centuries earlier and whose writings are lost. The *Almagest*, a Latin form of an Arabic rendering of the title *The Great Book*, is a condensation of 800 years of astronomical observations, and gives a sense of what the Greeks knew about astronomy. Ptolemy, a follower of Plato, undermined the physical reality of Aristotle's cosmology by adding epicycles into the description of perfectly circular planetary orbits. An epicycle – an idea taken from Apollonius of Perga of the third century BC – is a small additional circular orbit somehow described on the main circular orbit. It can have no physical meaning but is a way of ensuring that the model works mathematically. The addition of any number of epicycles ensures that the observed motion of a planet can always be described by circles. A cheat in other words.

Ptolemy never claimed his model as anything but a mathematical (or Platonic) description. His system used different formulae for calculating the position of each planet. In some ways it was barely more than tables of processed data, and not always very accurate data at that. There is no deep unification in Ptolemy's system, something we expect of a modern scientific theory. It is said that there were even epicycles on the epicycles, though there seems to be no evidence that this is true. In the thirteenth century, the astronomer and king of Spain, Alfonso X, is reported to have said of epicycles that if he had been at the Creation he might have given better advice. Although Aristotle's system is even weaker at describing the observed phenomena than Ptolemy's system, Aristotle's does at least have the advantage of possessing physical reality. By the sixteenth century it was clear that Ptolemy's great work was not all that it was hoped it would be.

The Church blessed the search for an improved cosmology that might establish a more reliable calendar. The obvious place to look for fresh ideas was to explore other newly rediscovered ancient writers. The Polish astronomer and cleric Nicolaus Copernicus (1473–1543) appeared to have found inspiration in Aristarchus from the third century BC, whose ideas (the original texts are lost) are preserved in the writings of Archimedes (c.287–c.212 BC). Aristarchus was the first person to argue for a sun-centred cosmos. He was even aware that a moving earth tells us that the stars must be a long way away, given that they do not appear to move. In ordinary life, when we move around objects that are near to us, we are aware that they change their spatial relationship to each other. This phenomenon is called parallax: simply an acknowledgement that there is a shift in perspective when we move between things. In the Aristotelian model of the cosmos there is no parallax between the earth and the stars because both are fixed: the earth unmoving at the centre of the universe, and the stars pinned to an outer moving celestial sphere some distance beyond the sun and planets. Any theory that has an earth that moves must account for the fact that the stars appear to be held in a fixed pattern (the constellations) that circles the earth every 24 hours. The fact is that there *is* parallax between the earth and the stars, but because the stars are so very far away they *appear* not to move. The tiny change of perspective is so difficult to measure that stellar parallax was not observed until the nineteenth century, when there were telescopes sufficiently powerful to make the sensitive measurements required. For many centuries, most thinkers took Aristarchus' argument that all stars are far distant as a reason to discount his sun-centred theory rather than as support of it.

Copernicus, who knew Ptolemy's *Almagest* inside out, realised that he could make Ptolemy's earth-centred model simpler if he, too, placed the sun at the centre of the cosmos. His model, like that of Aristarchus, is not strictly a heliocentric one so much as a heliostatic one: the unmoving earth is replaced by an unmoving sun. Copernicus continued to believe that the spheres were made of crystal, but he reduced the number of them from about 80 in the Ptolemaic system (the number had grown over the years) to 34.

Copernicus knew of Aristarchus' heliocentric system and obliquely refers to it in a surviving manuscript, but for some reason he does not cite the passage in the printed edition of his great work: *De revolutionibus orbium coelestium*. It is possible that he came across the ideas via the works of Arab writers. He delayed publication of *De revolutionibus* until after his death. It is often said that he did so in order to protect himself from the wrath of the Church, but it seems that he delayed because he hoped to first find proof, and because he feared the reaction of colleagues. It also seems likely that he was too busy, since as well as being an astronomer and Catholic cleric, he was also a classicist, physician, diplomat, philosopher, translator, jurist and governor. Copernicus had no more idea than Aristarchus how to account for the apparent lack of motion of the earth. Nor, since there is no fixed point from which to judge up and down, could he explain why heavy things fall to the earth. Any new theory that replaced a static earth with a moving earth would need to account for why objects fall to earth, as Aristotle's description does. Copernicus posited the existence of an attractive force that anticipates gravity, but he wasn't able to work it into a theory that could make measurable predictions. His force was mystical: 'but a natural inclination, bestowed on the parts of the bodies by the Creator, in order to combine the parts in the form of a sphere and thus contribute to their unity and integrity'. Nor is it entirely clear that his system was any simpler or more accurate than Ptolemy's. In any event, when his work was published it met with almost no reaction and was not banned until 1616, over 70 years after it was first published. Rather than cause a revolution, Copernicus's ideas may well have disappeared without trace had Galileo not taken an interest.

*

It was known from the thirteenth century that lenses could make distant objects appear nearer, but there were no telescopes until the Dutch invented them in the seventeenth century: novelty items made for spying on people across the street. Galileo Galilei (1564–1642) made his first telescope out of a verbal description given to him of the Dutch invention, and although he soon made telescopes superior to any in Holland, even his improved arrangement of lenses produced only hazy impressions, a world away from the crystal-clear images of modern instruments. Galileo may have trained his telescope across the street, but he made history when he trained it on the heavens and made sense of what he saw there. The English astronomer Thomas Harriot (1560–1621) was probably the first person to use a telescope for astronomical purposes.[8] In 1609, and subsequently, he began to map the moon, but it was Galileo who first realised that the moon had mountains and valleys.

In Aristotle's cosmos, the sublunar world is where things become degraded: because this is where change happens. Far from being the centre of the cosmos, the earth was the bottom of the universe, the place to which earthly objects fell. This was a view that found its way into Christian theology, certainly from the time of St Augustine (AD 354–430). In his *Divine Comedy*, the Florentine poet Dante Alighieri (1265–1321) places hell at the centre of the universe with Satan at the absolute centre. Even in the seventeenth century during the Reformation, the earth was regarded by some as the most unworthy of all the planets. Humanism was a reaction against this dismal theology and an attempt to find a more elevated placing for man in the cosmos.

In Aristotle's cosmology the heavens are to be found at and beyond where the moon is, a region that is both unchanging and spotless, literally immaculate. In Christian theology Heaven is, of course, seen as the most worthy of all locations. When Galileo described a moon that has mountains on it, and a sun that is spotted, here is evidence that Aristotle's cosmology is flawed, or at least in want of further elaboration.

This was the moment we began to trust technology to extend the reach of our senses, and when we began to believe that the universe has many of the same qualities that are evident on earth, that the heavens are not separate.

On 7 January 1610 Galileo identified three 'stars' close to Jupiter. On subsequent nights he saw that they changed position relative to each other, ruling them out as fixed stars. On 10 January he discovered that one of them had disappeared.

---

[8] He may also have been the man who first introduced tobacco to the British Isles.

Galileo had discovered three of the moons of Jupiter, one of which was now hidden on Jupiter's far side. On 13 January he identified a fourth moon. In less than a week Galileo had collected the first convincing evidence that not all heavenly bodies orbit the earth as they should according to the Ptolemaic system. Later in the year, Galileo observed that Venus has phases like our moon does. The Copernican and Ptolemaic systems make different predictions about how these phases should look when observed from earth. Galileo's observations favoured a system in which Venus orbits the sun, not the earth. As Galileo continued to collect evidence, the Ptolemaic system began to fail.

The Church did not ignore Galileo's discoveries, but it did reject the Copernican model as an explanation. The Church favoured a different model that was also in agreement with the new discoveries.

Tycho Brahe (1546–1601) was a Danish nobleman, astronomer and astrologist, whose most significant contribution to the history of science was the accuracy of his astronomical observations. It was on the foundations of Brahe's observations that the German astronomer, mathematician and astrologer Johannes Kepler (1571–1630) discovered his eponymous laws of planetary motion. In his description, astronomical bodies execute elliptical orbits, something that Galileo was not prepared to accept. (Kepler's laws were confirmed later, once Newton's law of universal gravitation was in place.)

Tycho Brahe believed that the cosmos is earth-centred, and devised a model that protects this aspect of Ptolemy's model. It was also used to explain the observations that Galileo would make after Tycho's death. In Tycho's model (for some reason Tycho, like Galileo, is known by his first name) it is conceded that Venus, Saturn, and the other known planets revolve around the sun, but the sun continues to revolve around the fixed point of the earth. Mathematically, Copernicus's and Tycho's models are equivalent. In fact the Copernican system has the disadvantage that the supposed motion of the earth and stellar parallax need to be explained.

It was Galileo's assertion that the earth does in fact move that the Inquisition forced him to renounce in 1633, and to which he is, famously and apocryphally, said to have added in a whisper: 'And yet it still moves!' (*E pur si muove!*) Effectively, Galileo was forced to deny his new scientific method, which held that the more elegant mathematical symmetry of the Copernican system made it a truer system than Tycho's. Galileo's attempt to ascribe physical reality to the Copernican model pitted mathematical elegance against the authority of the Church (as vested in the Bible, and certain classical ideas the Church had ossified). Galileo may have been forced to back down but his direct appeal to mathematical elegance as a final authority set science on a new course.

Perhaps it does not seem so unreasonable that the Church judged this a step too far. In a way, the Church was only doing what science does, refusing to accept a new model until the new model clearly describes more phenomena, and for which there is experimental evidence. It takes a brave soul to challenge the authority of the Church, just as it takes a brave soul to challenge the authority of science: neither embrace innovation with open arms. The difference is that no matter how dogmatic the tendency of the scientific establishment, the methodology of science ensures that all theories are provisional, and all theories must ultimately be replaced by new theories if progress is to be made.

Out of fear of the Inquisition, scientific investigation ground to a halt in the Catholic world and moved to England and Holland. The Church might have put its faith in the Tychonic system, but in the everyday world the Copernican system was quietly taken up, notably by navigators, and for the entirely practical reason that it was easier to use. Why put the earth at the centre if the mathematical calculations produce the same results but are more straightforwardly computed when the sun is placed there? But what the Copernican system could not yet address was why the sun should now be what was fixed at the centre.

Modern science could be said to have begun in that year

1543 when Copernicus removed the earth from the centre of the universe and put the sun there. With this single act he set out a principle by which science has been guided ever since: that not only is mankind not at the physical centre of the universe it is not at the centre in any fashion, literally or metaphorically. What launched the scientific revolution was not the placing of the sun at the centre of the cosmos (from where, anyway, it is later removed) so much as the *removal* of the earth. It's not about us.

Going Through The Motions

Our curiosity depends upon a receding horizon.
                                        Adam Phillips

The material world is a place where there are things, and those things are in motion. For 2,000 years, motion was largely taken to be as Aristotle had described it. Aristotle set out an elaborate metaphysics of motion, but fundamentally he asserted that an object does not move unless it is pushed, and that heavier objects fall faster than lighter ones.

Galileo spent much of his life trying to describe motion in a new way. His first work was titled *De Motu* (On Motion), and his last book, *Discourses and Mathematical Demonstrations Relating to Two New Sciences* (published in 1638 in the Netherlands, and without a licence from the Inquisition), returned to the subject. Galileo revolutionised the idea of motion. He showed that, regardless of their mass, all objects dropped from the same height hit the ground at the same time, or at least they would in a vacuum. He almost certainly arrived at this knowledge by thought rather than experiment. The relationship between thinking about and measuring the world is close knit and subtle. The famous experiment in which different-sized cannon balls were supposedly dropped from the leaning tower of Pisa, did not actually take place. The fact that in Galileo's thought experiment the cannon balls fall to the ground at the same time only in a vacuum is effectively a Platonic idealisation of the world, a way of tapping into how things must be rather than how they appear to be in the corrupted world wc inhabit. But out of these idealised notions a theory is constructed that can then be tested. Experiments prove that the world really is like that and not, after all, as it appears. A genius like Galileo is sometimes so sure that the world must be as he has understood it intellectually that the experimental outcome is presumed. Galileo may not always have tested his theories, but lesser mortals have. In the absence of great genius science usually works in the other direction. Corrupt nature is observed and measured, and out of that knowledge idealised theory is constructed. The modern scientific method could be said to be an extension and melding of the Platonic and Aristotelian philosophies. The observation part of the process is what we have inherited from Aristotle and the idealised mathematical description from Plato, though there is a danger that with hindsight we make too great a distinction between the two philosophers, a distinction that was meaningless at the time.

It was Galileo, not Einstein, who first realised that all motion is relative. In another thought experiment, he imagined two boats each travelling at a steady speed on a perfectly flat and empty sea (a scenario that could only exist in a Platonic world). By thought alone he saw that it would be impossible as a passenger on either boat to tell where the true motion is, only the relative motion between the boats is apparent. There is no experiment that can be performed that will tell me if I am moving, if it is you who is moving in the other boat, or if we are both moving. We need a shoreline, or something fixed, to measure absolute motion against. The universe, however, does not have a shoreline, not even in the so-called fixed stars, which are not fixed at all, but only appear so because they are so far away. The best we can say about the motion of all the bodies in the universe is that they are seen to be moving relative to each other.

The Ptolemaic cosmos has a stationary earth at its centre, making it the shoreline of the universe from which it is possible to judge all motion. In Galileo's cosmos there is no still centre, and hence no shoreline. Indeed there can be no still point anywhere in a universe made out of things all moving relative to each other. No thing in the universe can ever be said to be truly at rest. For humans on earth, the appearance of rest is another compelling illusion.

Isaac Newton (1643–1727) formalised and developed Galileo's ideas into his three laws of motion. The first law tells us that in a frictionless world, things move uninterruptedly and forever until impeded by an outside force. This law is also called the law of inertia. The second law describes what happens if a force is applied to an object (it accelerates), and the third that all forces come in pairs: whenever a force is applied, an equal force springs into existence and pushes the other way. Aristotle had come close to understanding the principle of inertia. It was, in fact, his reaction against such Platonic idealisations as eternity, a vacuum and a perfectly frictionless surface that led him to make the opposite conclusion. He argued that there could be no motion at all in a vacuum and so concluded that there is no such thing as a vacuum. Over the next 2,000 years the principle of inertia would be discovered several times over, notably by the Chinese philsopher Mo-tzu in the third century and by Arab philosophers in the eleventh century, but this new understanding of motion did not take hold until it was expressed by Newton as part of his radical reconception of what we mean by physical reality. With his three laws, Newton set out a mathematical description of a physical world in which there are concepts like mass, velocity, acceleration and momentum. To this new world he introduced a new and particular kind of force, and which he described separately in his theory of universal gravitation. As Copernicus had done before him, Newton posited that there is some inclination inherent in all stuff that reaches across the emptiness of space and pulls things together. The difference, however, is that Newton found a way to describe the nature of this force in mathematics. In a single equation he shows that the strength of the gravitational force is directly related to the mass of the bodies that are drawn to each other, and falls off as the square of the distance between them. Between them, Galileo and Newton worked out how to relate mathematics and cognition. Nature is 'written in the language of mathematics,' Galileo once wrote, 'and its characters are triangles, circles and other geometric figures, without which it is humanly impossible to understand a word of it; without these one wanders about in a labyrinth.'

Can we conceive what the world looked like before Newton described it in terms of velocity, mass and gravity, when gravity was a mood and not a force? Newton's conceptual world has become so real to us that we may be cut off from it in ways we will never be able to quite grasp.[1]

Newton's gravitational force is no less mystical than the force posited by Copernicus, but we are willing to pay the entry price this time around – to suspend our disbelief – because the descriptive power of Newton's theory is so encompassing. Gravity unifies the heavens and earth. The same force that causes an apple to fall to the ground also impels the moon around the earth. In Aristotle's physics, there are different descriptions for different parts of the cosmos: his explanation of why the planets move is different from his explanation of why things move on earth. The French philosopher René Descartes (1596–1650) attempted to explain planetary motion as the result of vortices in some sort of fluid that permeated space. Newton offers a single description for large and small objects across the universe: not a local description but a universal one. He describes what the mystical force posited by Copernicus would need to look like – how it needs to behave mathematically – if we are to explain why we are unaware of

---

[1] The American literary critic Harold Bloom has argued that Shakespeare 'invented' us as modern human beings. Can we say what it meant to be human before Shakespeare created the language of modern sensibility? (*Shakespeare: The Invention of the Human*, Harold Bloom, 1998)

the earth's motion through space. Gravity keeps the things of the earth, including its atmosphere, bound together locally, as if it were a ship ploughing through a void. At all scales of size gravity holds the structures of the universe together, as planetary systems, galaxies, clusters of galaxies and clusters of clusters of galaxies.

Newton's laws of motion tell us why, once in motion, the planets remain in motion. The planets are closer to the Platonic world of ideals, moving as they do through frictionless space. On earth, it is less obvious that motion is as Newton describes it. Moving objects slow and come to a stop, since on earth there is friction: the motion of things is impeded and its true Platonic nature obscured.

In a purely mechanical worldview such a mystical force as gravity ought to be disallowed, but scientists are pragmatists. If the theory works well enough, some mysticism will serve, for the time being at least, in place of too much mystery. Descartes's attempt at a mechanical description to explain the motion of planets seems to be truer to the spirit of materialism, but Newton's explanation, though it rests on an immaterial force, has the decided benefits of universality and mathematical elegance. Newton's gravitational theory is known as the *universal* theory of gravity for good reason. Galileo and Newton removed the stillness that had existed at the centre of the old cosmos, replacing rest with relativity. The earth moves around the sun at about 30 kilometres per second, by which we mean: if we suppose the sun to be still, the earth moves relative to it at 30 kilometres per second. But the sun isn't still: it is, for example, moving relative to the centre of the galaxy. The solar system takes 225 to 250 million years to orbit the centre of the galaxy, travelling at supersonic speeds (217 kilometres per second). The Milky Way moves towards Andromeda at 88 kilometres per second. The Local Group moves around the centre of the Virgo supercluster at 600 kilometres per second. And the Virgo supercluster moves around a complex of galaxies called the Great Attractor. In the universe at large everything is in motion at all scales of size.

Stillness is an illusion. We measure motion by reference to axes of time and space that we take around with us. It was Descartes (1596–1650), observing a fly in a room, who realised that objects could be uniquely described in the world of space and time by reference to a set of coordinates: three of space and one of time. We ensure that steady motion is measured universally as the same motion by different observers by making a simple addition or subtraction that translates the motion from one frame of reference to another: me on the earth, say, and you somewhere in the far reaches of some other spiral arm of another galaxy. It would be sheer egotism to say that rest is where I have determined it to be. To declare the earth as the still centre of the universe could only be achieved by fiat.

Newton's universe of separate moving things is played out in a theatre framed by space and time that has no place of rest. Space and time are immutable, eternal and infinite. Space is infinite in extent and time is measured out as if by a pendulum that measures out arcs in space for all eternity. In an empty universe, the Newtonian worldview assures us, there would still be time and space. Emptiness would have meaning. There would always be time and space, even when there is nothing.

For several hundred years Newton's description of the universe worked very well. But then in the faster modern world we began to realise that even his theory breaks down. Newton's laws of motion hold for the kind of speeds we associate with everyday life, but if we take a less egocentric view we discover that those laws are not universal after all. They do not work at very high speeds. Newton's conception of motion, it turns out, only works for limited types of motion. It is an approximate description, as ultimately all scientific theories prove to be.

Sometimes a theory can be saved by modifying it; at other times, in order to describe the world more closely, the world has to be to described differently. This is one of those latter times.

Albert Einstein (1879–1955) reconceived the universe in order to make sense of new phenomena that Newtonian mechanics could not explain. Einstein's remedy is drastic. Time and space are not absolute as Newton conceived them, nor are they how we think we apprehend them. Einstein realised that there is something else that is more fundamental than space and time.

The word fundamental, like the word unique, is not meant to have a comparative or a superlative. There is no uniquer or uniquest. If something is fundamental then surely there can be no deeper understanding, nothing more fundamental, yet in scientific discourse the bottom is always being taken out from under the world. We can never be sure that what we take to be fundamental qualities of the world will stay fundamental for very long. Truth in science is always provisional. In fact science can sidestep the idea of truth altogether. There is only what is truer rather than what is true. And there always is something truer. Progress in science might even be understood as the certain knowledge that there is always some quality, as yet unsuspected, that is more fundamental.

Einstein came up with a new idea of what motion is. He realised that all motion is the same as the motion that light has. What this means takes some getting used to. We have such a strong idea of what we think motion is that to conceive it differently is almost beyond our imagining. We are so comfortable with Newton's idea that time and space are absolute and that things move relative to that fixed framework that Einstein's theory still shocks us over a hundred years later.

Einstein's famous theory, the one known as the special theory of relativity, first appeared in 1905 in a paper entitled 'On the Electrodynamics of Moving Bodies'. It was the German physicist Max Planck (1858–1947) who renamed the theory, though Einstein thought the word relativity was misleading and would have preferred the word invariance instead, a word that has the opposite meaning.

Einstein inherited from Galileo the principle of relativity and the Copernican idea embedded in it that reality must look the same for all observers moving at steady speeds. He was also inspired by the work of the Austrian philospher and physicist Ernst Mach (1838–1916). In a thought experiment, Mach had realised that the motion of a single object in an empty universe is meaningless (since motion only has meaning relative to the motion of something else). Mach's principle, still not clearly understood, suggests that the whole universe is involved in any motion. The universe pushes back whenever we apply a force to an object. Using this principle Einstein was able to account for a tiny wobble made by the earth, unaccounted for by Newtonian mechanics, and which is due to the presence of all that is in the universe beyond the planets and the sun. Mach's curious principle seems to imply that somehow the universe knows to make the earth wobble; indeed, that the whole universe knows when an apple falls to the ground. Whatever the implications, Mach freed Einstein from the idea that fixed axes of time and space are needed to describe motion. In Newton's conception of physical reality time and space make up the framework within which the drama of the universe unfolds. Liberated from such a conception, Einstein set about removing the universe to a different theatre.

It had been known that light has a finite speed for well over a century, though it wasn't until the end of the 1840s that the French physicist Armand Fizeau (1819–1896) made the first decent measurement. By 1862 the measurement was accurate to within 1 per cent. It was Einstein, however, who realised that the speed of light must also be the fastest possible speed in the universe. This single assertion undermines the Newtonian idea of relative motion. Einstein is telling us that the motion of light cannot be relative. No matter what motion anything has relative to light that motion can never be faster than the speed of light (because there is no speed faster than the speed of light). In the Newtonian world the relative motion between two light beams heading towards each other is predicted to be twice the speed of light. In Einstein's world the combined motion cannot be greater than the speed of light. Because of something peculiar about the nature of light, Einstein realised that motion cannot be as Newton described it. To say that nothing travels faster than light is just another way of saying that the motion of light is invariant or *not* relative.

In the Newtonian world we think we live in, we think we know how to tell how one motion differs from another, if for no other reason than the fact that we can see things moving relative to us at different speeds. We even think we know how to measure these relative speeds. Crashing head-on into a wall is half as bad as crashing head-on into a car coming towards us in the opposite direction at the same speed.

Einstein shows us that this is not how the world is: it is merely how the world appears to be at the low speeds we are used to in our everyday lives. Admittedly, Newton's description is a very good approximation. It's not that Newton's description doesn't work at speeds close to the speed of light. It doesn't work for any speeds. It's just that it's harder to tell at slower speeds. Ultimately, science is not about approximate measurement, it is about ever finer measurement. Sometimes to measure more finely is to measure differently. When objects are travelling at high speeds – those close to the speed of light – it becomes clear that relative speeds cannot be added together in the simple way we are used to. In order to explain why Newton breaks down, Einstein had to describe motion in a different way. In a unified description of the world, how could it be possible that there is one sort of motion that has a different nature from the motion of all other things? It isn't possible and so there must be something wrong with our conception of how we think the world is.

Einstein does away with the idea of relative speeds and replaces it with the deeper understanding that motion is not relative: it is invariant and it is the motion that light has. But how can this be? How could an ambling pedestrian have the same motion as the motion of light? It makes no sense. But that's because we think we know what motion is and are thinking in Newtonian terms. Einstein describes motion differently, and he does it by uniting time and space more deeply. Space and time are joined together, Einstein says, into a single four-dimensional reality called space-time, and it is in that extra-dimensional reality that all motion is seen to be the same. Motion only looks different to us because we don't experience that unity of time and space.

Einstein unhinges the absolute and eternal reality of Newton's space and time and replaces it with a new absolute: the unchanging nature of light. Time and space become relative qualities because the speed of light is not. One of the consequences of this insight is that in the new world as Einstein reveals it to be, clocks appear to tick more slowly for anyone who is moving relative to us (in the Newtonian world that we experience). Perhaps even more bizarrely the person we are observing thinks exactly the same: that it is *our* clock that is slowed. And this symmetry must be true if the Copernican principle that we all observe reality in the same way is to hold.

From our perspective, for a beam of light time is slowed down so dramatically that it doesn't tick at all. Light has no motion through time. All its motion is directed through space. From this understanding we can perhaps gain a glimmer of understanding that shows us how all other motion is the same as the motion of light: all other motion is motion that is to greater or lesser degrees directed away from motion in space to motion in time. The ambling pedestrian seems to be moving slowly compared to the speed of light because most of the motion of the pedestrian is in time rather than in space. When we look at the motion in space-time, we can see (or a mathematician could persuade us) that the motion of the pedestrian and of light itself are equivalent.

There is, however, a deep problem with Einstein's description. The theory doesn't explain why we actually experience time and space separately. It is hard to accept that we live in a four-dimensional world not the three dimensions of space and one of time. Might we experience the deeper unity if we were more conscious, or is there something missing from Einstein's description that makes it a mathematical rather than a physical description of reality? Platonic rather than Aristotelian? But we might also remember that our natural response to the world is to put ourselves at the centre of it, and to believe in the existence of our ego, and that doesn't seem to be how the world actually is either.

Einstein struggled to make his theory a general theory. The special theory accounts for steady motion (and can account for accelerated motion) but does not account for gravity. A simple (but impossible) thought experiment shows us that there is something about gravity that puts it at odds with what we know about light. If the sun were to be instantly removed from the universe, we would notice the loss immediately since, according to Newton, gravity acts instantaneously. But we would not *see* the removal of the sun for eight minutes, as this is how long it takes light to travel from the sun to here. This apparent paradox is in need of resolution.

The instant transference of information implied in Newton's theory of gravity is disallowed in Einstein's. Einstein would have to come up with a new theory of gravity. His break-through came when he realised that there is symmetry between acceleration and gravity. He called his insight 'the happiest thought of my life'. But his theory was hard won. Einstein, who never considered himself much of a mathematician, needed to learn some very difficult mathematics in which to write his theory. It took ten years. The equations are complex, unlike the memorably elegant equation $E = mc^2$ that mathematically underpins the special theory of relativity. The English writer C. P. Snow (1905–1980) once remarked that if it were not for Einstein we might still be waiting for the general theory. This is not a universally held view, and, in any case, how could we ever know?

The triumph of the general theory is that the mystical nature of Newton's gravitational force evaporates into the geometry of space-time. The presence of mass distorts space-time, and gravity *is* that distortion. A planetary orbit is simply the route carved out of space-time by the mass of the sun. The earth goes round the sun because the mass of the sun (and, to a lesser degree, of the earth) distorts space-time, and the earth moves along that distortion. If the sun were more massive, space would be curved more tightly, making for a shorter and closer orbit.

Around objects with relatively low masses, like our sun, the distortion of space-time is more apparent as a distortion of space. Around the more massive objects the universe has to offer, like a neutron star, it would also be apparent that time is dilated. Finally, Einstein takes the mysticism out of Newton's force. 'Matter tells space how to curve, and space tells matter how to move' is how John Wheeler elegantly summarised it. Or as the American physicist Michio Kaku (b.1947) puts it: 'In some sense, gravity does not exist; what moves the planets and stars is the distortion of space and time.' Mercury, the sun's innermost planet feels the sun's gravitational effect most strongly. A slight oddity in its observed motion due to its being in a strong gravitational field is explained by general relativity.

The effect of gravity isn't only astronomical: the slowing-down effect of gravity has been measured more mundanely. A clock at the top of a 25-metre-high tower on the Harvard campus was shown to run faster – by 1 second every 100 million years – than a clock at its base, where gravity is slightly stronger. The difference in the timekeeping abilities of the clocks is just as predicted by general relativity. Now that's a precise measurement.

If the sun were suddenly removed from Einstein's universe, space-time would unfold in a wave that travels towards us at the speed of light. We would feel the disappearance of the sun gravitationally at the same moment we see its disappearance. Both light and gravity are ways of transferring information and both are limited by the speed of light. There is a hint here that perhaps gravity and light are somehow connected, that they may even be aspects of something deeper that unites them. Einstein spent the second half of his life attempting to unify these two fundamental aspects of nature, a search that continues today.

*

Modern cosmology began with the general theory of relativity, but the mathematics is so complex that it wasn't initially clear how the theory was to be interpreted physically. Wondering how he might apply his theory to the universe as a whole, Einstein realised that he had to make some simplifying assumption about the nature of reality, about how the universe has to be. Einstein's assumption was that the universe should look the same when viewed from any vantage point; that is, it should be isotropic. This assumption is called the cosmological principle and is another affirmation of the Copernican idea that no one has a privileged position in the universe. No one is at its centre, literally or figuratively. In this case, Einstein turns the Copernican principle on its head: if nowhere is the centre of the universe and the universe looks the same from all vantage points, then this is equivalent to saying that everywhere is the centre. We are at the centre of the universe after all; but then so is everything else.

When the general theory was first published in 1915, there was much speculation about how far the universe extended. It was thought by many that perhaps the Milky Way was the whole universe. But from here on earth we see the Milky Way as a sort of road through the sky that the Romans called the *Via Lactea*.[2] If this was the whole universe it would look different to aliens not looking from our vantage point, and so, according to Einstein's cosmological principle, it cannot be the universe.

Einstein made a further simplifying assumption: that the contents of the isotropic universe are distributed smoothly. This is quite an assumption. Obviously the universe is not smooth and featureless at any of the orders of magnitude we have come across, but at the largest scales, beyond the structures formed by clusters of clusters of galaxies, there are reasons to suppose that Einstein's assumption is correct. It would seem though that there is a certain circularity of argument here that is hard to escape: out of the assumed smoothness of the universe, smoothness is discovered. Once more,

---

[2] 'Milky Way' is a translation of the Latin *Via Lactea*; indeed the word 'galaxy' comes from the Greek word for milk.

we must put aside such philosophical quibbles. The cosmological principle allows scientists to apply general relativity to the universe at large and out of it comes a deeper understanding that is affirmed by experiment and evidenced by technological progress. Shut up and calculate!

Within months of the publication of the general theory, the German physicist Karl Schwarzchild (1873–1916) found a mathematical solution to the general theory that predicted the existence of black holes, though they would not be named as such until 1967. At first, the idea that there could be such strange objects in the universe was strongly resisted, even by Einstein himself. Gradually, these superdense bodies would come to be recognised as one of the most important features of the universe.

By 1917 Einstein and others realised that the equations of general relativity did not describe a static universe. Einstein was worried that gravity would cause a universe that was not stationary to contract. He fixed the situation by adding an arbitrary term he called the cosmological constant (not to be confused with the cosmological principle) to keep the universe steady. Adding additional and arbitrary terms to a theory is not something scientists do lightly. Ptolemy's theory was supported by the repeated addition of epicycles, an ad hoc solution. An arrangement of epicycles could always be found to explain new data relating to planetary motion, but at the cost of making the theory mathematically cumbersome and without physical reality. In the modern scientific method the physical interpretation of a theory is paramount. Without a physical interpretation science is reduced to mathematical abstraction. Later, Einstein called the addition of his cosmological constant the biggest blunder of his life.

In 1922, the Russian mathematician Alexander Friedman (1888–1925) realised that without the cosmological constant the equations of general relativity actually describe an *expanding* universe. Again, it was Einstein who first resisted the idea that such an interpretation of his own equations could have any physical reality. The Belgian priest and astronomer George

Lemaître (1894–1966), however, decided to take this solution at face value. He argued that if the universe is expanding then it must have expanded from something smaller. Taking this idea to its logical conclusion, he conjectured that time and space can be unwound to a place where the whole universe was in one place at one time: that the universe had a beginning before which there was no time and space.

Lemaître's reasoning could hardly have been more suggestive, made explicit when he described the universe as a 'cosmic egg exploding at the moment of creation'. That the universe had a beginning was not, for many scientists, a happy thought: barely disguised Christian dogma, in fact, presented by a Catholic priest posing as an astronomer. But Lemaître had come to an accommodation with his dual life: 'There are two ways of arriving at the truth,' he once said. 'I decided to follow both of them.'

Until the 1920s science held, if unwittingly, the belief common to some Eastern religions that the universe is eternal, infinite and uncaused, in need of neither an act of creation nor a creation story. Newton's universe never ends and never began, consonant with the notion of an eternal and unchanging God. It is God's laws that Newton sought to understand. God created the world and nature's laws, but the universe itself exists as God does, eternally. To look out into Newton's universe is to look back eternally to ever more distant objects.

The English astrophysicist Sir Arthur Eddington (1882–1944), putting aside his own personal repulsion at the idea that there had been a moment of creation, spotted in Lemaître's paper of 1927 a possible explanation of a troubling observation made two years later by the American astronomer Edwin Hubble (1889–1953). One of the assumptions of science is that the universe is consistent. Whatever we understand to be the properties of, say, the elements from close study of them on earth, we believe those elements have the same properties when we come across them in other parts of the universe. When elements burn, the light that is emitted produces a distinctive pattern of colours when viewed at the atomic level using a process called spectroscopy. The spectrum of colours can be used to identify each element uniquely. In 1929 Edwin Hubble found the distinctive spectrum of hydrogen in a distant star, except that its spectrum was shifted over, as if all the colours were more red than they should be. Assuming that this is hydrogen we are looking at, and since we believe that hydrogen is the same all over the universe, this phenomenon is in want of an explanation.

The Doppler effect provides a simple explanation. The Austrian physicist Christian Doppler (1803–1853) was the first to account for an effect that we all notice almost every day in modern life. It is most apparent today when an ambulance passes by at high speed. The sound starts off higher than the true pitch of the siren and reduces to a lower pitch as the ambulance passes by. It sounds as if the pitch drops suddenly, but in fact it is the intensity of the sound that drops and the pitch actually falls steadily throughout. Like parallax, the Doppler effect is the outcome of a changing relationship between observer and what is observed.[3] A similar thing is happening when we see the spectra shifted over to the red part of the spectrum. A red shift tells us that the colours appear to have less energy. The Doppler effect tells us that that appearance of less energy corresponds to an object that is receding from us. The drop of intensity is an effect of the motion, not something inherent in the object itself. In other words we are looking at the same hydrogen, but it is hydrogen that is receding from us. The explanation is simple but the implication is momentous, and had already been predicted by Lemaître.

Lemaître had made the bold prediction that the most distant bodies would seem to be moving away from us in an expanding universe, and will appear to be moving away faster, the further away they are. Distant bodies are being carried away from each

---

[3] If you don't hear the Doppler effect when the ambulance approaches, that's because the ambulance is heading straight for you. It might be a good idea to step aside.

other, not by any inherent motion, but by the expansion of space itself. Hubble's 1929 observation and subsequent observations supported this interpretation. Nearly all the distant objects in the night sky appear to be moving away from us, and the best explanation is not that they all happen to be moving away by coincidence but because it is space itself that is expanding and carrying them away.[4] One of the first observations of these mysterious red-shifted spectral lines was of Andromeda's spectrum. The shift of the lines over to the red end of the spectrum could be explained if Andromeda is assumed to be moving away from us at what was then the fastest astronomical speed yet measured. Repeated observation of red-shifted astronomical objects (once thought to be stars) was taken to be further evidence of both an expanding universe and of galaxies other than our own. Andromeda and all of these distant retreating astronomical objects were upgraded to galaxies, and an extraordinary unification of nature was effected: knowledge of the atomic world was used to understand the universe at astronomical sizes. As it had when Galileo first lifted his telescope to the heavens, the universe reconfigured itself. Not only can we understand the whole universe by looking out to its largest extent, for the first time we begin to understand the universe at large by looking at the universe more finely at its smallest dimensions.

With the cosmological constant removed, a new solution – the so-called Big Bang theory – is plucked from the mathematics of general relativity, though once again after initial resistance from the theory's author. Now when we look out into space we look back, not into Newton's eternity, but to the beginning of the universe. Looking out to the horizon is to run the history of the universe back in time. The slow-moving matter that we see

---

[4] It is a little more complicated than this. Galaxies do have their own motion, and some of them are moving towards us and some of them away. The point is that, overall, the expansion of space is also taking them away from us. The further away a galaxy is, the clearer this is, because here the galaxy's own motion becomes less significant than the expansion of space itself.

around us moves faster and becomes more radiation-like (more and more like light) the smaller the universe becomes. When the universe began there was only high-energy light, out of which all the matter that is in the universe was created. The universe is light that has evolved.

Some of the light from a time soon after the Big Bang reaches us today, having travelled across the universe since the beginning of time. It reaches us in a form of radiation called cosmic microwave background (CMB), discovered in 1965 and the single strongest piece of experimental evidence that the Big Bang happened. The radiation was first mapped by NASA's Cosmic Background Explorer (COBE): launched in 1989, the results were announced to the world on 23 April 1992. A new mapping was made by NASA's Wilkinson Microwave Anisotropy Probe (WMAP) in the early years of the new millennium. It provides both the most detailed picture we have of how the universe looked soon after the Big Bang and the most accurate dating yet of the age of the universe. On 11 February 2003 NASA proclaimed the universe to be 13.7 billion years old plus or minus 200 million years. The map has become more detailed in recent years. A three-year map was released on 17 March 2006, and a five-year map on 28 February 2008. This latest data means that the universe can be dated even more accurately. The universe is now said to be 13.73 billion years old plus or minus 120 million years.

Einstein reconceived what motion is in order to unify the motion of all things. The nature of light shows us how to make that unification. He also exposed some deep connection between light and gravity, those two means by which we see, and ultimately apprehend, the universe. The Big Bang tells us that the universe began as a ball of light 13.7 billion years ago. All those years later, we are aware today that the universe has complex things in it, including human beings who have discovered that everything was once undifferentiated radiation. But if everything was once light, it seems reasonable to ask how some of that light became stuff and how that stuff became us. And science does ask, and has answers to, these questions.

So far in our story we have seen into the universe by means of light and by the presence of gravity. That there is a connection between light and matter is continually hinted at as science progresses.

By means of light and gravity we can conceive the universe as a container with an edge in which there are moving objects processing information. The edge of the universe is a horizon. It is as far as we can see to: what is called the *visible* universe. There is every reason to suppose that there is much more of the universe – perhaps an infinite universe – beyond that horizon, where the laws of nature might be quite different. Metaphorically, we also see the universe by trying to understand it. When we consider the universe, that metaphorical ability to see is the same as our literal ability to see. If we want to see further, we must think more deeply about the nature of light and gravity.

The Other Way Out

But confronted with the statement that atoms were 'so small they were no longer small', one lost all sense of proportion, because 'no longer small' was tantamount to 'immense'; and that last step to the atom ultimately proved, without exaggeration, to be a fateful one. For at the moment of the final division, the final miniaturisation of matter, suddenly the whole cosmos opened up.

Thomas Mann, *The Magic Mountain*

Tiny stars are a constant reminder that the universe is very large. The vastness of the heavens has been celebrated in culture for as long as there has been culture. The heavens are where the gods live, to whom we have looked for signs and meaning.

But there is nothing visible to the naked eye to remind us that the universe is also very small. It is not at all obvious that there is a realm of tininess in balance with the heavens. One of the first to explore this new territory was the English scientist and architect Robert Hooke (1635–1703). His book *Micrographia* (published 1665) was a bestseller, and was noted for its detailed engravings (some of which may have been made by Christopher Wren), particularly that of a flea, an illustration that opened out to four times the size of the book. In his diary, Samuel Pepys

(1633–1703) wrote that he stayed up to 2 a.m. he was so engrossed, remarking that it was 'the most ingenious book that ever I read in my life'. Only with the invention of the microscope had it become possible to see that there is a corresponding world of meaning to be found in the very small. Though the microscope and the telescope were Dutch inventions of the early years of the seventeenth century, Hooke's revelations came over half a century after Galileo first lifted his telescope to the sky.

The vastness (if that is the right word) of the smallness of these worlds is apparent in the almost boyish desire of those early pioneers to count what they saw, so excited were they by Nature's profligacy. Hooke estimated that there are 1,259,712,000 chambers in a square inch of cork, early evidence of the cellular structure of living forms. The Dutch scientist Anton van Leeuwenhoek (1632–1723), whose microscopes were not improved on for over 150 years, counted 8,280,000 animalcules in a drop of water.

The seventeenth century saw the opening up of the universe at the largest and smallest dimensions, but it wasn't until the twentieth century that the astonishing smallness of the world figured in our physical understanding of the universe as a whole.

We have no name for the tiny character of the universe ('faeryland' will surely not do), and no language by which to navigate small spaces. We reach out to the stars, but where do we reach to when we reach out to small things? Inwards? Downwards? Below?

We have a notion that some day we might travel the universe to see for ourselves that it is as we have described it, but how could we ever explore the world of atoms and subatomic particles? There is no access to the microscopic (unless we imagine ourselves as smaller versions of ourselves, like Alice shrunk to get through a tinier-sized garden door), except as passive observers using our ingenuity artificially to extend our ability to see.

It is easy to think that the universe is all about largeness. After all, what is the height of a human compared with the distance away of even our nearest stellar neighbour? A human being does not seem to be much bigger than nothing. We more readily accommodate the idea of a place that contains everything than that place which, symmetrically, ought to contain an idea of what nothingness is. Space contains everything, but where is nothingness? It appears tantalisingly nearby. We understand that to approach nothingness we must be getting closer when things get smaller and smaller, just as we believe we can only encompass the universe by looking at larger and larger objects until there is nothing else left to see. We can imagine holding an object that shrinks until within our own fingers it eventually arrives at that place where nothing is. But where is that place? It is not a location or a destination, and yet it seems to be an inherent destination in all locations.

Materialists believe that the world is made of something and that that something can be measured and described. Materialists are forced to say: since there is something, that something must be made of something else that is smaller. If a material description of the world must ultimately include a description of what the nothing is that something comes from, then it is inevitable that a study must be made of ever smaller things. In physics that study is of atoms and their constituent parts.

If the largest structures in the visible universe can be encompassed by a modest number of steps of increasing order of size, then correspondingly and logically, it should be possible to search out the smallest structures in the universe by making steps of decreasing order of size. We can descend (if it is descend) into the world of small things, shrinking from objects a metre long to a tenth of a metre (10 centimetres) to a hundredth of a metre (1 centimetre), a thousandth of a metre (1 millimetre), and so on. There are structures of smallness to be found that are as mysterious, maybe even more mysterious, than those of the universe at large. And every step we take brings us closer to the greatest mystery of the material world: what is meant by nothing at all.

100–10 centimetres (10^{0}–10^{-1} metres)

The smallest adult human was Lucia Zarate (1864–1890), a

Mexican woman exhibited at Barnum's Circus, who at her tallest was 50.8 centimetres (1 foot 8 inches). A human foetus at full term is, on average, 51.3 centimetres from crown to heel. Before the age of 20 weeks foetuses are measured from crown to rump, since at this time the legs are curled up against the torso.

10–1 centimetres (10^{-1}–10^{-2} metres)

At 14 weeks a human foetus is, on average, 8.7 centimetres from crown to rump.

The smallest bird in the world today is the bee hummingbird at 5 centimetres long. Its nest is only 3 centimetres wide. The smallest mammal is either the bumblebee bat from Thailand or the Etruscan pygmy shrew, depending on how smallest is defined. The bumblebee bat is 3 to 4 centimetres in length and weighs about 2 grams. The Etruscan pygmy shrew is around 3.6 centimetres long and weighs 1.3 grams. The bumblebee bat, however, has the smallest skull size (1.1 centimetres) of any mammal.

10–1 millimetres (10^{-2}–10^{-3} metres)

The smallest fish are members of genus *Paedocypris* from Indonesia and are 7.9 millimetres long. Even smaller, at between 6.2 and 7.3 millimetres, are the male anglerfish of the species *Photocorynus spiniceps*, but the females are much larger.

1–0.1 millimetre (10^{-3}–10^{-4} metres)

A pinhead is about 1 millimetre across. Most of us have no problem seeing pinhead-size objects as small as a thousandth part of our metre rule, but only with excellent eyesight can humans discern something that is a tenth of a millimetre (10^{-4} metres) long, four orders of magnitude smaller than we are ourselves. Most mites and ticks are of this size: almost microscopic. The Broad mite (*Polyphagotarsonemus latus*) is less than 0.2 millimetres long.

0.1–0.01 millimetres (10^{-4}–10^{-5} metres)

The smallest insect eggs cross the boundary between what is discernible and indiscernible to the naked eye. The eggs of the parasitic fly *Zenillia pullata* can be as small as 0.02 millimetres long.

Cells, out of which all animal and plant life is constructed, and one of the basic structures to which all life may be reduced, are typically in this range of sizes. The cells of our body are as small relative to our own size as a mountain is large.

0.01–0.001 millimetres (10^{-5}–10^{-6} metres)

The smallest single-celled organisms – blue-green algae and bacteria – lie in this range of sizes. That these organisms are the oldest living creatures and also our oldest living relatives is a modern evolutionary discovery that bolsters the materialist belief that a fundamental understanding of nature arises out of the study of the smallest things.

Viruses are to be found at the bottom of this range. They are lengths of DNA that are typically around 0.001 millimetres (10^{-6} metres) long. Though viruses are technically the smallest forms of life, they cannot live independently of the larger life-forms that host them.

1,000–100 nanometres[1] (10^{-6}–10^{-7} metres)

The smallest recorded living organism is the *Nanoarchaeum*. It lives deep under the ocean in the severe conditions that

---

[1] A nanometre is a billionth of a metre, 10^{-9} metres.

surround hydrothermal vents, where the water is boiling. *Nanoarchaea* are typically 400 nanometres in length.

100–10 nanometres (10^{-7}–10^{-8} metres)

Since 1996 some scientists claim to have found living organisms, called nanobes, that are even smaller than *Nanoarchaea*. But others say that these 20-nanometre-long structures are merely crystal growths.

What life is at its most fundamental is constantly being revised downwards. Evolution has traced back the common ancestors of all living beings to algae and bacteria; to go further takes us into the world of complex molecules. Molecular biology is perhaps the most fertile area of research in evolutionary biology today. Most molecular biologists favour the idea that life first arose out of self-organising molecules, and the first possible candidates are beginning to be identified.

It appears to be increasingly apparent that life is not a hard boundary between the animate and the inanimate but something diffuse like the edge of the solar system or, indeed, the edge of the universe. Life begins to look like it may be some arbitrary label we impose on a phenomenon that is not entirely discrete, and whose meaning only gradually emerges out of an evolutionary process that must, ultimately, merge with whatever descriptions we have for the smallest structures in the universe.

10–1 nanometres (10^{-8}–10^{-9} metres)

A beard grows a few nanometres in the time taken to raise a razor to the skin.

Buckminsterfullerene – a man-made football-shaped sphere made out of 60 carbon atoms (C_{60}) – is important in the history of the development of nanotechnology. It is named after the American utopian and designer Buckminster Fuller (1895–1983) because of his association with the geodesic dome, a complex spherical or near-spherical structure that has the property of being much stronger overall than its constituent parts. Buckminsterfullerene is a geodesic sphere made out of 20 hexagonal and 12 pentagonal planes of carbon atoms. When compressed it is twice as hard as diamond, another form of pure carbon. Its highly symmetrical shape gives it many interesting chemical properties. A single C_{60} molecule can fit into a gap 1 nanometre wide.

A nanotube is another type of fullerene. It is a cylindrical carbon structure a few nanometres wide, but perhaps a few millimetres in length, in other words, typically a million times longer than it is wide. These man-made carbon forms have many applications in the emerging science of nanotechnology and in electronics.

Nanotechnology is the science that builds machines and structures out of molecule-sized parts, currently used in, say, computer chips or, more prosaically, in the making of stain-resistant clothing and to enhance the colloidal properties of suntan lotion. In the future, nanotechnology may find ways of carrying drugs around the body to specific sites. The American engineer Eric Drexler (b.1955), one of the founding fathers of nanotechnology, has predicted that one day nanomachines smaller than bacteria could be sent into space to build new materials, molecule by molecule, out of raw material already existing there: conquering the smallness of space becoming one of the means, perhaps, by which we conquer outer space.

Molecules are made up of atoms, and typically fall into this range of sizes. Long molecules, called polymers, can, in theory, be long enough to stretch into several of the orders of size preceding this one, but even polymers are not more than a few atoms wide. The most famous long molecule is the DNA molecule. The double-helix of the DNA molecule is 2 nanometres wide.

1–0.1 nanometres (10^{-9}–10^{-10} metres)

In this size range are the atoms from which molecules are built. The largest atom, that of caesium, is 0.546 nanometres wide, in the middle of this range of sizes. The smallest atom is the hydrogen atom at 0.106 nanometres across, only just above the bottom of the range.

All large-scale, or macroscopic, matter is made of molecules and all molecules are made out of atoms of the 94 different naturally occurring elements. There are other elements that only exist in laboratories. We can reduce nature to this modest number of differences: the elements that are familiarly arranged in the periodic table.

Atoms were meant to be the last word in a physical description of matter, the word 'atom' meaning indivisible. We now know that they are far from indivisible, though we also know that they not easily divided. Atoms make a strong barrier not easily breached that runs between the familiar world of 'things' and the curious world that lies beyond.

Atomism was a philosophy first propounded by Leucippus and his pupil Democritus before 400 BC and based on the idea that the universe was made of small things that are imperceptible, indestructible, indivisible, eternal and uncreated. Such a way of conceiving the world was not taken up again until the beginning of the nineteenth century when the English scientist John Dalton (1766–1844) realised that in some reactions chemicals combine in whole number amounts by weight, which suggested to him that there might be a smallest amount of each substance. So in modern times, atoms arose out of chemistry rather than physics. Dalton's idea, however, remained hypothetical for a further hundred years. Even in the late nineteenth century the physicist Ernst Mach, whose understanding of the relativity of motion influenced Einstein, was of the opinion that atoms could never be perceived by the senses, and so must be taken to be purely theoretical entities rather than physical ones.

Atomism, the philosophy that there are indivisible things, is troubling. How can there ever be an end to smallness in a material world? Any particle with extension in space cannot be both indivisible and fundamental. Anything that is extended is capable of being labelled: this area A and that area B. Such a thing must either be capable of being divided – even if we don't know how to do it – or we are forced to say by fiat that the thing is made of some material that can't be divided and so is not itself fundamental. Physicists in search of the fundamental parts of matter say that such parts must be without structure. Modern-day scientists have come to the same philosophical conclusion that the ancient critics of atomism arrived at, that if there are elementary particles they must somehow reveal themselves as point-like, that is, not extended. Anything that is extended must have structure. That elementary matter can have no extension in space led the ancients to rule out atomism. Though modern science comes to the same conclusion about what elementary matter must look like, it has, nevertheless, gone in search of this mysterious stuff.

Historically, as soon as it became clear that atoms have size, they were done for, and the search continued for the 'atoms' of atoms, for elementary particles, as they became known.

But what sort of existence, we might ask, can we ascribe to something that has no extension in space? Must *these* things, as Mach's atoms once were, now be taken to be purely mathematical entities without physical meaning? And how are we to find such mysterious substances in a net that is meant to be sifting ever smaller objects? Our elementary particles are bound to fall through if they have no extension in space.

Whatever the philosophical reality of atoms, it was becoming increasingly clear from experimental evidence that even atoms – whether they actually exist or not – are made of something else.

Below 10^{-10} metres

According to the British science writer John Gribbin (b.1946), it was the invention of an improved vacuum pump in the mid-nineteenth century that led to 'the biggest revolution in the history of science'. Perfect vacuums do not exist either in nature or in the laboratory. Nature comes closer. Most of outer space is more nearly a vacuum than any vacuum mankind has manufactured in a laboratory. But those first vacuum pumps did create a new condition on earth: an approximation of nothingness. Out of that new condition came a new understanding of the nature of reality.

An electric current flowing through a vacuum can be made to produce glowing rays that travel in straight lines and cast a shadow. These beams, first discovered in the 1860s, were named cathode rays. For a time it was thought that maybe they were a form of light, until it was discovered that they travel slower than light does. In the last years of the nineteenth century the English physicist J. J. Thomson (1856–1940) was able to show that these rays are made up of streams of electrons, particles even smaller than atoms. In 1899 he measured the charge on an individual electron and showed that it had a mass about 1/2,000 the mass of a hydrogen atom. Whatever atoms are, here was evidence that atoms are indeed made of something much smaller.

The French physicist Henri Becquerel (1852–1908) accidentally discovered radiation in 1896 while investigating various phosphorescent materials, such as uranium, that had accumulated since the time of his grandfather, who had made a study of such phenomena. What were first called Becquerel rays were later shown to be two types of radiation – streams of charged particles – named alpha and beta rays. Because these rays are energetic enough to take part in nuclear reactions, and because they occur naturally, they were crucially important in the early history of particle physics. An explanation of alpha and beta radiation contains a large part of the subsequent history of particle physics.

In 1909 the New Zealand-born physicist Ernest Rutherford (1871–1937) supervised an experiment using alpha rays that would reveal a recognisably modern picture of the atom. In the experiment, a thin sheet of gold foil is bombarded with alpha rays. Gold foil is used because it can be made very fine, essentially a few atoms in thickness. It was expected that the alpha rays would pass through the foil and be deflected through an array of angles as predicted by the then current model of the structure of an atom. In 1904, J. J. Thomson had put forward what was later called, though not by him, the plum pudding model of the atom. Negative electric charges are embedded in a cloud of positive charge, as if they were plums in a pudding.[2] The result of Rutherford's experiment showed that this model needed to be revised. Most of the alpha particles that make up the rays were not deflected at all as they passed through the foil, and what is even more striking, about 1 in 8,000 particles was deflected by over 90°. It has been said that it was like firing bullets at a sheet of paper and finding the occasional bullet coming straight back towards the gun. By 1912 Rutherford correctly interpreted the result, that occasionally an alpha particle – itself positively charged – makes a head-on collision with the concentrated positively charged centre of the atom, what was later called its nucleus. In this new model the tiny nucleus of the atom is surrounded by electrons that somehow move around it, like planets around the sun. How the electrons move about the nucleus would be one of the key discoveries of quantum physics. In 1912 such an explanation was missing.

A typical atom is a fraction of a nanometre (less than a billionth of a metre) across, but Rutherford revealed that to find the nucleus of an atom it is necessary to plunge through four more orders of magnitude. A typical atomic nucleus is around 10^{-14} metres across,[3] more than 10,000 times smaller than the atom itself. Coming across the nucleus inside an atom has been likened to finding a pea suspended in the space of

---

[2] Plum pudding is misleading. What is now often forgotten about this early model is that Thomson gave the electric charges the freedom to move about, a freedom forbidden to plums in puddings.

[3] A nucleus can range in size from 10^{-15} metres across (the nucleus of the hydrogen atom, which is a single proton), to around 1.5×10^{-14} metres for the nucleus of a large atom like that of uranium.

a cathedral, but such homely analogies, powerful as they are, can also be distracting. The atom is not a cathedral, nor the nucleus a pea; they are not scaled down things of our local world. In science, it is important not to trade understanding for the egocentricity of home, not to grant precedence to peas and cathedrals. 'Analogies prove nothing,' Freud once wrote, 'but they can make us feel more at home.' Home, by definition, is somewhere nearby. Science, by methodology, means to make home ever more inclusive: home as the universe.

Putting aside any philosophical conundrums, the search for the elementary material of the universe appeared to be going well. By 1932, the 94 differences of the elements were reduced to just three. The different atoms of nature are arrangements of protons and neutrons (as the nucleus) and electrons. In turn, however, protons and neutrons were discovered to have size (each are in the region of 10^{-15} metres across) and so were ruled out too as fundamental particles.

By the middle of the twentieth century, it had been discovered that protons and neutrons are made of yet smaller matter, called quarks, a highbrow name taken from James Joyce's largely unread novel *Finnegans Wake* ('Three quarks for Muster Mark'). It was the American particle physicist Murray Gell-Mann (b.1929) – a forbiddingly patrician and precise physicist, said on occasion to have corrected people on the pronunciation of their own names – who conceived the recherché theory that led to the search for, and discovery of, these mysterious particles. And it was he who named them. With a lighter touch more suited to his populist nature, Richard Feynman,[4] who was developing a theory along similar lines, proposed that the particle be called a parton, after Dolly Parton.[5]

---

[4] For relaxation he visited topless bars, where he drank 7UP and, if inspired, would jot down his thoughts on a napkin.

[5] The American country music singer Dolly Parton (b.1946) may be historically unique for having lent both parts of her name to science. The first cloned sheep, Dolly, cloned from breast flesh, was also named for her.

Quarks and electrons have a property that distinguishes them from other small particles. They are described as being *at most* 10^{-18} metres (a billionth of a billionth of a metre) wide, a size we arrive at only by plunging down several more orders of magnitude from the size of a proton or neutron. Quarks and electrons are claimed as the elementary particles for which we have been searching. But if elementary particles are to have no extension in space, what could they look like. And if they have no extension how can we also say that electrons and quarks are at most 10^{-18} metres wide? The clue lies in that 'at most'.

Light on the Matter

Kick at the rock, Sam Johnson, break your bones:
But cloudy, cloudy is the stuff of stones.
Richard Wilbur, 'Epistemology'

Our understanding of the large-scale world deepened because
we became more sensitive to light: we constructed better telescopes and so extended the reach of our eyes, and constructed
better theories and so extended the reach of our brains. Something similar seems to be happening at the other end of the
world. Microscopes are another way of extending our ability
to see.

Clearly, light is fundamental to our description of the material world, but we have not asked, as yet, what a material description of light itself might look like. Special relativity unifies
energy and matter through the speed of light, and general
relativity points towards a possible unification of light and
gravity. We know that gravity is the geometry of space-time,
but what is light? Our common sense impression of the world
is that there is matter to make a visible world, and then there
is the invisible world of light. If a unified description of nature
is to be made, somehow we have to understand how the invisible world of light becomes the visible world of matter. We

must resolve the paradox that what we see is seen by something that is itself invisible.

Aristotle thought that the eyes generate the light we see by. The first studies of light were related to the nature of the eyes themselves and how they see. The subject is called optics. Leonardo da Vinci may have been one of the first to note the diffraction of light into colours as seen, say, on a feather, or for us today on the close-knit tracks of a CD. Newton heard his contemporary the polymath Robert Hooke 'speaking of an odd straying of light caused in its passage neare the edge of a Rasor, knife or other opake body', a phenomenon that was not understood for at least another century. Newton had become interested in the nature of light after he bought a prism from a travelling fair. He used it to break visible light up into its constituent colours and then wondered, symmetrically (a useful scientific ploy), whether the colours could be recombined to make invisible light again. He had to wait until the fair came around again before he could buy another prism to put his theory to the test. Although Newton, in the late seventeenth century, made important discoveries about the nature of light, his conception of it as a stream of 'fiery particles' could not explain the strange effect that Hooke had observed around the edge of a knife. Newton's contemporary Christiaan Huygens had come up with a wave theory of light. One of the problems with a wave theory of light is that the waves need to propagate through something. Huygens proposed that they move through a sort of superfine jelly called aether. (The substance out of which the Aristotelian heavens were said to be fashioned was given the same name, but it is, of course, not the same stuff.) Newton's objection to aether was that it would need to fill the whole of space, and would slow down the planets. The aether problem was to dog science for a long time to come. Newton's fame, and his support of a particle theory of light, set back the study of light by more than a century.

In the first years of the nineteenth century the prodigious English scientist and Egyptologist Thomas Young (1773–1829) offered the first plausible explanation of the phenomenon that Hooke had observed. Young supposedly kept a notebook from the age of two, had read the Bible twice by the age of four, was reading Newton aged seven, and by the age of sixteen read Latin, Greek, French, Italian, Hebrew, Chaldean, Syriac, Samarian, Arabic, Persian, Turkish and Ethiopic. He was known as Phenomenon Young at Cambridge. He devised an experiment to try and prove that the diffraction of light around sharp edges is a phenomenon best explained if light is thought of as waves and not particles. His simple experiment of 1801 is one of the most famous and easiest to reproduce in the history of science. Through a pair of slits about a millionth of a metre wide (about the width a razor blade would make) he shone a bright light. On to a screen beyond, a series of bands of darkness and brightness shows where the light that has passed through the slits has combined together, as if it were water waves, cancelling out where the troughs and peaks meet and reinforcing where trough meets trough or peak meets peak. The effect is called interference. In 1802 William Wollaston discovered that light passed through a prism and looked at under a microscope also reveals dark and bright lines. Though it wasn't known at the time, he was looking at the sun's signature: what elements the sun is fashioned out of. A similar observation made later in the nineteenth century led to the discovery of helium, an element then unknown on earth, and named for *helios*, the Greek word for the sun. The sun was once thought to be the same as the earth only hotter. The Presocratic philosopher Anaxagoras (*c.*500 BC–428 BC) thought that meteorites were material thrown out from the sun. Even in the 1920s it was thought that the sun was only 5 per cent hydrogen. We now know that it is almost all hydrogen with some helium and small amounts of a few other elements.

So long as the benefits outweigh the cost, science can be surprisingly flexible in what it will allow. Newton's invisible gravity was accepted (against some resistance, of course) because the gravitational theory unites so much of nature. If a new theory together with the necessary evil of some new unexplained substance leads to greater explanatory power – that is, explains a great deal of previously unexplained phenomena – then the necessary evil is borne. Aether at least has the advantage of being material, even if it is not known what the material is.

Taking Young seriously meant taking the aether problem seriously, and aether took some getting used to. Young's work was ridiculed for decades, but gradually the explanatory power of his wave description began to outweigh the discomfort of having to accept the existence of an unexplained and unobserved substance for the waves to travel through.

In 1887, the American scientists Albert Michelson (1852–1931) and Edward Morley (1838–1923) devised an experiment that was, in theory, accurate enough to detect the wind that should result from the passage of the earth through the aether as the earth makes its orbit around the sun. Despite their best endeavours they found no evidence of such an aether wind. Their experiment is an experiment in a particular sense of the word. They tested the idea that there is an aether wind and *falsified* it within the error limits of the apparatus they used.

The Austrian-born British philosopher Karl Popper (1902–1994) turned the idea of what constitutes scientific proof on its head. A scientific theory can never be proved absolutely, he realised, but it must always be falsifiable. Newtonian mechanics, for instance, is falsifiable because it asserts that all motion is relative. The theory was falsified as soon as it became clear that a better theory can be constructed on the basis that the motion of light is *not* relative. Ptolemy's description of planetary motion, on the other hand, is not a properly scientific theory because the epicycles used to bolster the theory of circular planetary motion make the theory unfalsifiable. No matter what planetary motion is observed, it is possible to construct a description of that motion using as many epicycles as it takes to fit the data.

More usually we think of an experiment as finding evidence rather than confirming an absence, but the Michelson – Morley experiment is still an experiment in Popperian terms. The experiment leaves open the possibility that aether could still exist if finer measurements are made, but then that is true of all finer measurements: that reality might look quite differently if we look a little more closely.

It was Einstein (who else?) who would prove that aether does not exist, and through pure intellect rather than through physical measurement of the world. In his special theory of relativity, Einstein makes the assumption that light does not have a relative speed. His theory is effectively based on a belief that the aether cannot exist. Light does not move relative to anything, and that must include aether. The success of his theory meant that the question of what light propagates in was quietly dropped. How light could be a wave but not a wave *in* anything began to seem like a purely philosophical question. (Shut up and calculate!) In any case, by then the nature of light was about to be understood in a new way. Two tributaries of the river of science were about to meet up in a major unification of nature.

*

The Greek philosopher Thales knew that sparks (what we now call static electricity) could be produced when amber is rubbed against certain materials, and that amber buttons attracted hair as if by the power of some force. Certain rocks were also found to possess this ability to attract. The discovery of these two natural attractive forces produced by what we now know to be electricity and magnetism was the first tiny inkling that there might be a connection between them. In the mid-eighteenth century the American polymath and founding father Benjamin Franklin (1706–1790) noticed that electricity and magnetism each comes in two forms, which he called negative and positive. This ad hoc property is posited to account for the observation that some charges repel and some attract. Knowing that like charges repel and opposite charges attract is not an explanation: it is an assertion that that is what charged things do. It is, however, the beginning of an explanation. Sorting things out into types is what science does first. It separates out and labels, then tries to join disparate phenomena together in unexpected ways.

In the early nineteenth century, the Danish physicist and chemist Hans Christian Oersted (1777–1851) showed that a compass point moves towards an electric current flowing through a coil of wire, the first demonstration that an electric current produces magnetism. A decade later, the English physicist and chemist Michael Faraday (1791–1867) took up this idea and wondered, symmetrically (that important notion again), if a magnet moved through a coil of wire might produce an electric current, and so discovered what is called inductance. He also discovered that electricity can affect polarised light, which suggested to him that there is a connection between electricity and light.

Faraday, who was the son of a blacksmith, first read about electricity when, during the course of his job as a bookbinder, he read an article on the subject in *Encyclopaedia Britannica*. His break came when the English chemist and physicist Humphry Davy (1778–1829) took him on as an assistant after another assistant had been dismissed for drinking. If magnetism and electricity have a particle nature, Faraday wondered how the particles knew how to move. The same problem is inherent in Newton's gravitational theory: how does one body know how to affect another from a distance across empty space? Because of the success of the law of universal gravitation that problem was, over time, forgotten about.

Faraday imagined that each electrical and magnetic particle is surrounded by 'an atmosphere of force', a sort of condition of space that he later called a field. Faraday, like Einstein, was a strong adherent of the idea that nature is unified. Though Faraday's grasp of mathematics was notoriously poor (nor was he much interested in it), he invented the concept of a 'field' (fundamental to how we understand the particle nature of the universe) in order to unite these phenomena in an explanation. It is this field of influence that tells the particle how to move. It can be envisaged as arrows, at every point in space, that indicate the direction of the force. Whether it is purely a mathematical description or something real, it is hard to say. The field doesn't explain in a deep way why the particles at a distance know which way to go; it is an ad hoc addition, as are aether and charge, that makes the explanation work. But by giving the field some properties, we find that we can explain other phenomena that would otherwise have remained unexplained. The field description is so successful at this that we gradually grow comfortable with the idea of fields, as we did with gravity, and eventually even grant the field a physical presence in the world. The epicycle was an ad hoc addition, but in Ptolemy's cosmology every finer measurement of planetary motion required the addition of yet another epicycle, whereas Faraday's field idea uses a single ad hoc concept to unify a multiplicity of phenomena.

Nature encourages us to believe in our ad hoc phenomena. Just as we see apples fall and so believe in Newton's gravity, even though we didn't know what it was until Einstein turned it into the geometry of space-time, so we see iron filings align themselves along a magnetic field and begin to believe that a magnetic field must exist. The field is our explanation of how a force can jump across space, as gravity does across a gravitational field in Newton's description. The field effectively *is* the force.

The discoveries of Oersted and Faraday begin to suggest what we now know to be true, that electricity and magnetism are inextricably entwined. A stationary charged particle is surrounded by an electric field, and a moving charged particle produces an electric field and a magnetic field. So, if a charged particle moves through a magnetic field, the magnetic field changes because the moving charged particle also produces its own magnetic field. In turn, that changing magnetic field produces an electric field that causes the electric field of the moving charged particle to change, which once more causes the magnetic field to change, and so on. This reinforcement is what electromagnetic radiation is. The cycle of reinforcement of electric and magnetic fields moves at the fastest speed possible, the speed of light.

Changing electric and magnetic fields cannot be described independently of each other. A changing magnetic field produces an electric field and a changing electric field generates a magnetic field. When they reinforce each other, electromagnetism – or light – is the result.

In the same way, an accelerating electric charge also produces electromagnetic radiation: a static charge produces an electric field; a moving charge produces an electric field and a magnetic field; an accelerating charge produces a changing electric field and a changing magnetic field. But since a changing magnetic field also produces an electric field, once again the fields reinforce each other as electromagnetic radiation.

Light, by which we once meant visible light, as from the sun or a candle, turns out to be part of a continuous range, or spectrum, of radiation called electromagnetic radiation. Visible light is just a tiny part of that spectrum, which we sense with our eyes as being separate from other parts of the spectrum: infrared radiation is felt as heat, ultraviolet radiation tans the skin, X-rays destroy cells. By naming them, we have separated them out as separate phenomena with different properties, but underlying this separation is a continuity of energy that moves from radio waves with the lowest energy to gamma rays with the highest. Nature doesn't know that we have named little parts of it. As far as she is concerned, light is one continuous form. Casually, scientists use the word 'light' to mean any part of the electromagnetic spectrum.

The tiny part of the spectrum that is visible light is divided further into a range that we see as colours. The ends of the whole electromagnetic spectrum are described as the red end and the blue end, as if the spectrum were an extension of that small section called visible light. Radio waves are not red but they are found beyond the red end of the visible part of the spectrum; similarly X-rays and gamma rays are found beyond the blue end.

In 1861 the eccentric Scottish mathematician and physicist

James Clerk Maxwell (1831–1879), known as Dafty at university, published a paper that contained four equations that fully described the mathematics of electromagnetic radiation and that showed that such radiation would travel at the speed of light. Maxwell guessed that electromagnetic radiation is the same thing as light, but at that time there was no experimental evidence. The German physicist Heinrich Hertz (1857–1894) provided the physical evidence when he produced the first radio waves and microwaves and showed that all electromagnetic waves travel at the speed of light. Although Maxwell's equations are less well known than Newton's laws of motion and Einstein's relativistic equations, they are just as significant in the history of science. These seemingly abstract equations, together with Hertz's physical evidence, unified electricity, magnetism and optics into a single description. It is only because we tell the story of this synthesis historically that we are inclined to think that electricity and magnetism are the real parts of something more tenuous called light. Because we came to describe electricity and magnetism first, they are more embedded in scientific theory and so we grant them the greater reality. We see the parts more clearly than we see the whole, since the whole is what is always being pointed at, something away on the horizon and in the future. That light is both less tangible and the deeper reality is a concept that is hard to grasp. Scientific investigation puts a strange cast on reality. The deeper understanding is always the most provisional because it is the most novel, the latest arrival, least tested, the most hypothetical. Magnetism and electricity are discovered to be aspects of something more symmetrical, and that something more symmetrical is light.

Maxwell's equations appear to contain a fatal flaw: they only seemed to work for stationary observers. There is a way out of the dilemma if it is assumed that the speed of light is invariant. But this goes against any understanding we have of motion as described by Newtonian mechanics. Newton's laws of motion tell us that all motion is relative. How can one sort of motion be invariant? It was Einstein's faith that

Maxwell's equations must be correct and Newton's mechanics wrong that encouraged him to take this implication at face value and run with it: that the motion of light is special. Out of it came his special theory of relativity and a deeper understanding of how electric and magnetic fields are related. According to special relativity, these fields must be manifestations of the same thing because viewers moving in different frames of reference will see what appears to be an electric field in one frame of reference as a magnetic field in another.

In that same miraculous year of 1905, when Einstein published his paper on special relativity, he threw in two other seminal papers.

In one, he came up with conclusive evidence that atoms actually exist (though by then it was clear that they were not elementary particles, and were made of yet smaller particles that may or may not be said to exist). He published a paper that made sense of a phenomenon called Brownian motion. This peculiar phenomenon, which had gone unexplained for almost 80 years, was first described by the Scottish botanist Robert Brown (1773–1858) as a dance that pollen performs on the surface of water. Einstein realised that the motion could be seen as the perpetual impact on the pollen of individual water molecules, and though he only tentatively made this suggestion – 'I could not form a judgment on the question' – Einstein's tentativeness often masks certainty. In any case, from this point onwards atoms became entities with physical reality, fundamental in twentieth century chemistry and physics.[1]

In the third paper of 1905 he offered an explanation of a hitherto unexplained phenomenon called the photoelectric

---

[1] Einstein worked out the mathematics of the random walk that the pollen makes on the surface of the water, which turns out to be the same kind of walk that sunlight makes between molecules of air in the earth's atmosphere. Because blue light is more easily scattered than other colours is why the sky is blue, and blue in all directions. The reason had been known for some time, but Einstein was the first to explain the phenomenon mathematically.

effect. This paper turned out to be a foundation stone of quantum physics.

Hertz had discovered an effect he could not explain: that light at ultraviolet wavelengths can produce a spark if shone on a metal plate. This is the photoelectric effect. The problem is that even very dim ultraviolet light produces the effect but differently coloured lights do not, no matter how energetic they are. This effect cannot be explained using classical descriptions of light, according to which light of any colour that is energetic enough ought to produce the sparking.

Einstein solved the problem by borrowing a recent idea that the German physicist Max Planck (1858–1947) had used in order to explain another troubling phenomenon in classical physics called the blackbody problem. It had been observed that, when substances are heated, most of the light emitted makes a peak of the same shape. The peak frequency may vary for different substances, but the shape of the peak stays the same. Why there is a peak at all could not be explained out of the classical understanding of light as a smooth wave. Planck had a solution: he decided to treat light in the same way that heat is understood. Heat is the measure of the energy of a particle, and temperature is the average energy of many individual particles. Planck decided to treat light as if it were lumpy like heat. He never meant to suggest that light is actually made out of lumps; in fact he made it quite clear that he believed this not to be the case and that his assumption was merely a mathematical device. Out of this assumption, however, he was able to solve a tiresome problem. Planck did not know why radiation should be treated this way; he just knew that he could explain the observed phenomenon if it were treated so. Einstein borrowed the same idea, but whereas Planck had used the idea mathematically, Einstein treated the idea physically, just as he had decided to take the implications of Maxwell's equations physically. He decided to believe that these lumps of energy do actually exist.

Instead of a continous wave, Einstein said that light should be seen as packets of waves, effectively giving it a particle-like behaviour. The classical wave theory of light says that brighter light is more energetic and so brighter light should increase the photoelectric effect. In reality this does not happen. If light of a particular frequency (or colour) does not produce sparking, then no matter how much brighter the light is made the sparking will never occur. Conversely, if the light is of a colour that makes the sparking happen, then the sparking will continue to happen no matter how dim the light is. In Einstein's new description, the energy of light is described in a different way. Einstein's light is broken up into packages (effectively particles) called quanta, from the Latin *quantum*, 'how much'. It was Planck who calculated how large these quanta should be, even if he had not at that time appreciated that quanta actually exist. Just as the large value of the speed of light shapes the landscape of the macroscopic world, so does a tiny number called Planck's constant shape the landscape of the subatomic world. The size of each quanta of energy is the frequency of light multiplied by Planck's constant, which has a value of 6.626×10^{-34} joule seconds.[2]

In 1926 the American chemist Gilbert Lewis (1875–1946) named these quanta of light 'photons'. If the quantum package, or photon, has sufficient energy, then an electron is ejected from one of the atoms that makes up the surface of the metal: the ejected electrons are what we register as sparking. In Einstein's new description, no matter how few particles make up the light – no matter how dim it is – the effect happens. But if the quantum doesn't have enough energy, then no matter how many particles are shone on the metal – no matter how bright the light is – the sparking will never occur.

With this paper, Einstein effectively invented quantum physics. Almost a decade later, the Danish physicist Niels Bohr (1885–1962) used the same idea to explain another hitherto unexplained phenomenon. The atoms of a heated gas emit light in well-defined colours in spectral lines. We have seen

---

[2] A standard unit used to measure energy spread over time.

that the spectral pattern is a unique signature that can be used to identify every element, but why there are spectral lines is beyond the explanatory power of classical physics. To solve the problem, Bohr proposed a new theory of what the atom might look like physically. It was effectively a first stab at a quantum description of the atom, and a refinement of the Rutherford model, that had in turn replaced Thomson's plum pudding model. It quickly became apparent that Rutherford's planetary model of the atom couldn't be how atoms actually are. An orbiting electron is an accelerating charged particle emitting electromagnetic radiation. An electron emitting electromagnetic radiation is an electron that is losing energy. It has been shown that an orbiting electron would fall into the nucleus in less than a ten-billionth of a second.

In Bohr's description, electrons whiz around the nucleus in well-defined orbits. The electrons can only move between these different orbits if they possess well-defined amounts, that is, quanta, of energy. In this way, the spectral lines are evidence of the movement of electrons between these discrete orbits. The difference in energy that an electron has in one state as opposed to another is seen as an emission of light that makes up the spectral line. The fixed size of these lumps of emitted energy is evidence, as Einstein assumed, that quanta of light really do exist. We have already seen that it was Bohr's model that allowed cosmologists to work out the make-up of distant stars and to interpret an observation made by Hubble as evidence of an expanding universe. Why the electrons are confined as Bohr states is another of those assumptions we have to accept, and are willing to accept because more phenomena are explained if we do.

Quantum physics changes the way we describe light, and since it is light that we use to describe reality, it also changes the way we describe reality. A world that appeared to be sinuous, reveals itself, when looked at closely, to be grainy, in the same way as our technological world which was once smoothly analogue is now pixilated and digital. From a distance we can't tell the difference, but when we get in close we can.

In the quantum description of light, we see things because the tiny particles of light named photons hit them. If the particles are sufficiently small and there are enough of them, we can imagine them getting into all the crevices of whatever it is that we are looking at and revealing enough detail to give the impression that we are looking at a coherent whole, rather as a film, made as it is of 24 flickering framed images every second, gives the illusion of real life. To see things more clearly would, in this scenario, require the existence of smaller particles.

Common sense tells us that, if we only see something by throwing small objects at it, then no matter how gently we lob those particles the object we mean to look at must always be disturbed in some way. There is no escaping the fact that if we see with particles that possess energy, they must transfer some of that energy on to the object being viewed. Common sense also leads us to the conclusion that this way of viewing the world puts a limit on how precisely it can be viewed. No matter how finely we try to view the world, we must always disturb it.

It might seem reasonable enough to assume that an observable world exists beyond these limits even if we cannot see it and therefore know de facto that it is there. But a quantum description of the world is the place where reasonableness and common sense must be set aside. Richard Feynman said of the quantum physical description of reality that 'nobody understands' it, which ought to give us some encouragement. It is as if the energy of the particles acts as a veil pulled across the world, behind which reality takes the opportunity to be something else altogether.

We have an idea what we think reality looks like: a world of separate moving things that we describe using the familiar Newtonian (or classical) concepts of position, velocity, mass, energy and time. In classical mechanics, knowing the momentum (measured by multiplying mass and velocity together) and the position of a particle fully describes its motion. In theory we can calculate where the particle was in the past and where it will be in the future. However, in 1927 the German physicist Werner Heisenberg (1901–1976) propounded his famous Uncertainty Principle, which states that momentum and position *cannot* both be precisely measured at the same time, so denying a complete classical description of nature.

Newton's laws describe a mechanical world of discrete objects moving in time and space. In such a world, at every moment, each object with mass has a position and a velocity that can be measured relative to some frame of reference. In theory, once we know the motion of an isolated body at a single instant of time we know where to find that body at every point from the past and into the future. Its motion is described by two pieces of information, the position of the object and its momentum. And if the body is not isolated, then again in theory (in practice it turns out rather harder to do) so long as we know the motion at some instant of each body in the system then we can similarly describe the system as a whole. At the beginning of the nineteenth century the French mathematician and astronomer Pierre-Simon de Laplace (1749–1827) wrote of an intelligence that might comprehend all 'the movements of the greatest bodies of the universe and those of the lightest atoms . . . For such an intelligence nothing would be uncertain; and the future, like the past, would be open to its eyes.' Heisenberg tells us that the world is not like this. Of course we immediately want to ask why, as we did as children. Why is the Heisenberg principle true? Unfortunately, we must first accept that this is just how the world is, and then out of this new description we look for predictions that can be tested experimentally. It is the confirmation of these predictions by experiment that encourages us to accept that Heisenberg's principle provides a better description (in that more phenomena are encompassed) than the description we had before.

Heisenberg's principle tells us that the more precisely we measure the momentum of a particle, the less precisely will we know where the particle is, and vice versa. We can imprecisely know the particle's momentum and its position, but never precisely know *both* quantities at the same time. If we want to know the *precise* momentum of a particle, then we must give up any idea that we know where the particle is. It's not that the particle could be anywhere so much as the fact that the idea of location becomes meaningless. And conversely, if we know *exactly* where to locate a particle, we must give up the idea that we can know its velocity: the idea of velocity and hence momentum becomes meaningless. This is our new picture of what reality looks like. It is very different from a classical view of the world, which assumes that we can measure the motion of an object, constructed out of the knowledge of its position and momentum, ever more precisely.

Momentum and position are approximations of the world of larger-sized things that we call the classical world. The reality of the quantum world is different. Before we make a measurement, the particle exists in such a way that it has no momentum or location. Once we make a measurement we extract a certain amount of information from the quantum world, which we see as an uncertain description of the particle's momentum and its position in a classical world. The qualities that Newton discovered (or is that invented?) and used to describe his classical world no longer exist in the quantum world that lies behind it.

If we want to describe the particle in classical terms, we discover that such a description must always be incomplete. Before it is observed, a particle cannot be said to be anywhere. Location is a property of the macroscopic world. In its quantum state, before we make a measurement, a particle exists with the possibility of being in different places. Only after the measurement is made will the particle reveal some information about its uncertain position. Heisenberg's Uncertainty Principle tells us that the classical world cannot be precisely known. Precise measurement is not ruled out, but a precise measurement does rule out a complete classical description of nature.

Quantum mechanics only makes sense (if it can be said to

make any sense) if we understand that the process of measurement creates the appearance of what we call classical, or observable, reality. Before they are measured, things are not things, but exist in a state of potentiality that is describable by a wave of mathematical probability. The wave unfolds in time and only means anything when we extract information out of it, that is, ask what some aspect of a particle's observable nature looks like at that instant. Reality isn't unknowable but it is uncertain. Heisenberg actually used the untranslatable word *anschaulichen*, which more accurately means indeterminacy or indeterminability. In its quantum state the whereabouts of an electron is not so much uncertain as indeterminable.

We cannot know with precision the answer to the two questions that would otherwise have determined the classical nature of reality. We give precedence to the classical world because we are sure that the world is made up of things that move in space. Quantum physics tells us that that world is an illusion constructed out of the partial information we extract from a deeper reality. The physical world of things is simply the form that this partial information takes. It is the sense we make out of it.

Now we can begin to understand the strange relationship between elementary particles and size. Knowledge of the size of a particle is, as we have seen, merely a trade-off with the other aspect of a particle's classical nature, its momentum. A particle with low energy only appears to possess extension in space. In the classical world everything has low momentum and everything appears to have extension in space. This is an illusion that emerges out of a deeper quantum reality. This appearance of extension is not the particle's true size: size is merely a quality that is most apparent at the low energies of the ordinary world. At high energies an elementary particle becomes point-like. But even the point-like nature of an elementary particle is not its 'true' size either: it is what it looks like when we choose to measure the particle's energy.

How the mathematical formalism of quantum physics was to be interpreted physically proved to be as much of an intellectual challenge as those first attempts to find physical interpretations of Einstein's equations of general relativity had been. But first came the challenge of setting out the mathematical formalism, and it was Heisenberg who made the first attempt. His breakthrough came when he found himself on the island of Heligoland in the North Sea, escaping a severe bout of hayfever. He realised that he could make use of a mathematical entity called a matrix that had previously been restricted to the world of pure mathematics. (Einstein had also used an innovative mathematical language in which to write his equations of general relativity.) A lot was at stake, philosophically, in these early days of quantum physics. The pioneers had divided into two camps: followers of Bohr who emphasised the gappiness of the quantum leap, and followers of Einstein who emphasised the quantum world's double nature, as both wave and particle. Heisenberg's matrices put him in the Bohr camp. The Austrian physicist Erwin Schrödinger (1887–1961) was, however, so repelled aesthetically by Heisenberg's formalism that he called it 'crap' (or, presumably, whatever the equivalent word is in German). He determined to come up with his own description, and so booked himself into a hotel in Switzerland for two weeks, taking with him his mistress and two pearls (to put into his ears to block out any background noise). Schrödinger's wave equation of 1926 betrayed his own taste for a wave description of reality, and his dislike of the discontinuity inherent in the idea of quanta. He was led to his equation after pursuing an analogy with a vibrating violin string. In 1924 the aristocratic French physicist Louis, seventh duc de Broglie (1892–1987) had shown that all particles could also be described as waves, even a ball can be said to have a wave nature. As ever, nature reveals itself to be profoundly symmetrical. The German physicist Max Born (1882–1970) argued that Schrödinger's wave is not the particle itself but a measure of the probability attached to its particle nature. For a time there was a quiet battle between the Schrödinger and Heisenberg formalisms, even though mathematically they are equivalent.

In turn, the English physicist Paul Dirac (1902–1984) was

not impressed with Schrödinger's equation. If Schrödinger found Heisenberg's mathematics ugly, Dirac thought the same of Schrödinger's. In 1928 Dirac came up with an elegant description of a certain class of particles that also encompassed Einstein's theory of special relativity: the first attempt at a description that encompasses both quantum physics and relativity. Dirac essentially began the process of turning quantum physics into a field theory, which it still is, though much elaborated on since Dirac's time. It is as fields and particles that the physical world is described today.

Dirac's equation predicted the existence of particles with negative energy, though at the time no one was prepared to grant this solution to the equation any physical meaning. Heisenberg said of Dirac's equation that it was 'the saddest chapter in theoretical physics'. But particles with negative energy were discovered. The first evidence of so-called antimatter was made in 1932 when a particle was detected that was identical to an electron except that when it met an electron it annihilated it. This antielectron is otherwise known as a positron. It was subsequently discovered that all elementary particles have antiparticle partners.

Heisenberg's limit on measurement imposes a trade-off not just between momentum and position but also between energy and time, another pair of quantities that would provide a complete classical description in Newton's world. According to Heisenberg's principle, pairs of particles made of matter and antimatter are allowed to exist at very high energies as long as they don't exist for very long. Because overall they cancel each other out, they cannot be said to exist in the classical world, nor to break any of the laws of classical physics. Their existence is effectively borrowed and paid back out of nothingness, or out of what scientists call the vacuum (not to be confused with the vacuum we remember from school that attempts to create an absence of matter with the aid of a pump). Such particles can borrow time and travel back in it, or travel faster than the speed of light without breaking Einstein's law.

The vacuum is only nothing when looked at from a distance. The closer we get, the more energetic does nothingness appear to be. Aristotle had argued that nature abhors a vacuum; in fact he believed that there was no such thing as utter emptiness, and it seems he was right.

Energetic particles come into and out of existence without cause. They pop into and out of existence at random, beyond the edge of the world of causes: the classical world with which we are familiar of large, slow-moving and low-energy things. It requires increasing amounts of energy to peek into the tremendous energy of nothingness. The more energy we put into the vacuum the more particles causelessly pop out of it. The impossibility of absolute nothingness guarantees that, rather than fade into emptiness, the world does the opposite when looked at closely: it becomes more and more energetic. There is no such thing as empty space. Outer space appears to be mostly empty, but close up the tiniest part of space reveals itself to be less and less empty. Emptiness is the quality space appears to have at larger sizes.

All the visible matter in the world is made up of just four particles: two sorts of quark (called up and down), the electron and a particle associated with the electron called a neutrino. Unfortunately, in order to account for these four particles, the existence of hundreds of other particles is required (plus their corresponding antiparticles). All of these particles have been found by inserting large amounts of energy into the vacuum, from which they can be briefly brought into some kind of existence. There are so many of these particles – seen as evanescent spikes of energy in particle accelerators – that they have been collectively named the particle zoo. The Italian physicist Enrico Fermi (1901–1954) was heard to say to a questioning student: 'Young man, if I could remember the names of these particles, I would have become a botanist.'

This profligacy is troubling, and the search for elegant laws underpinning nature has seemingly taken a reversal. The most convincing evidence that current field and particle descriptions are on to something comes from the fact that these theories are the most accurately tested theories in the history of science, more accurately tested, even, than Einstein's relativity theories. Quantum field descriptions have been tested to within an accuracy of one part in a billion, as if the distance between New York and LA were measured to within the thickness of a hair. The inelegance of these quantum field theories, collectively called the Standard Model, is, however, profoundly troubling. The theoretical physicist Thomas Kibble (b.1932) has gone so far as to say of the Standard Model that it is 'such an extraordinary ad hoc and ugly theory it [is] clearly nonsense'. And even its fans admit that it is cobbled together. If the Standard Model is itself ugly, it is at least suggestive of deeper symmetries.

Though there are only two quarks that go into the making of visible matter, there are three such pairs in the Standard Model, called up and down, charmed and strange, and top and bottom. These six qualities are called flavours. The very names of these properties tell of their arbitrariness. One quark is no more charmed than the other or more up or more bottom. What these qualities might look like in the physical world it is impossible to say. In a way this is just as true of the ad hoc labels we have already come across, such as the 'positive' and 'negative' labels that Benjamin Franklin gave to the charge on what was later discovered to be the electron. Positive and negative only have meaning in so far as we see one sort of magnetic or electrically charged object repel or attract another. They are arbitrary labels used to distinguish the phenomena. 'Top' and 'up' are just two more arbitrary labels needed to distinguish other phenomena found in the world, the only difference being that these phenomena are too far removed from our local world to reveal any physical expression there. Certain particles have another ad hoc property called spin, introduced by the Austrian physicist Wolfgang Pauli (1900–1958) to account for the fact that only certain numbers of electrons can exist in an atom at any one particular energy level. The spin allows us to say that electrons are confined to shells of different energies. The difference between the energy levels of electrons confined to different shells is measured in quanta. We have seen that it is the movement between shells and the subsequent release of quanta of energy that explains the uniquely defined electromagnetic spectra of elements. This arbitrary property of spin might be said to exist in the hinterland between a property like charge, which we have a feeling for in our everyday life, and the up and down quality of quarks which has no counterpart. It is possible to transfer spin from particles in the quantum world on to an object in our local world, say on to a ball. So for this reason we might choose to agree that quantum spin is more meaningful than whatever strange and bottom mean. And yet, because all properties in the quantum world are quantised, that is, exist in discrete amounts, it becomes hard to say even what quantised spin or quantised charge might mean. Worse, it turns out that certain particles have half amounts of spin, which drains the word of whatever literal meaning we might have thought we could attach to it. The further quantum field theories delve in order to search out the ultimate symmetries of nature, the more abstract are the qualities that must be attached to the world as we find it to be. Not only have scientists identified six quarks, but each of them can exist in one of three different forms, or, as these forms are arbitrarily labelled, three different colours. So now there are effectively 18 different sorts of quark.

But there is, fortunately, something encouraging in the way they are manifested in the mathematics: the six quarks have the three symmetrical qualities we have already mentioned, and, further, the many particles that can be made out of quarks are always made out of either three quarks or two quarks. The proton, for example, is made up of two up quarks and one down quark and the neutron is one up quark and two down quarks. The triplets and pairs of the quantum world keep appearing in the Standard Model, though nobody knows why. The hope is that it indicates that there are undiscovered symmetries the discovery of which will simplify our description. Science could hardly have progressed if it didn't believe, in the words of John Wheeler, that 'everything important is, at bottom, utterly simple'. If science has a credo, then this would be its first line. Scientists believe that the universe is unified, and that the unification can be described by elegant mathematics. The Standard Model is far from being utterly simple, but it does intimate that there may be some underlying simplicity we do not yet understand.

The particle zoo is pulled out of the vacuum using high-energy particle accelerators. Now that we have done away with the idea that high-energy particles have anything that can meaningfully be called size, we can refine our notion of what it is to see. Instead of seeing things by throwing ever smaller particles at them, we see into the world of small things by conjuring up ever more energetic particles out of the vacuum. Particle accelerators do what microscopes did at larger sizes: they allow us to peek into the domain of the smallest things. A particle accelerator is the ultimate boy toy. All it does is smash energetic particles head-on in order to entice particles that are even more energetic to come into existence from out of nothingness. In 2009, with luck (and after a failed first attempt in 2008), a new particle accelerator will go online. The Large Hadron Collider (LHC) is a series of circular tunnels 100 metres underground, the longest of which is 27 kilometres in circumference, running under the border between France and Switzerland. This largest tunnel is made up of 9,300 superconducting magnets, each weighing several tons and chilled to the temperature of deep space. The LHC has the capability of accelerating particles up to 99.9999991 per cent of the speed of light and smashing them into each other. With machines like this, no wonder Murray Gell-Mann said: 'If a child grows up to be a scientist, he [*sic*] finds that he is paid to play all day the most exciting game ever devised by mankind.'

Obviously there must be some benefit to be derived from this multiplicity of particle production or we would have given up on this approach long ago. The benefit is, as ever, that the Standard Model manages to unify a great deal of phenomena – much of nature indeed – into a single description: a description entirely made out of particles. Einstein thought that nature might be unified if light (electromagnetic radiation) and gravity could be unified. Unfortunately, investigation of the world of small things requires two additional forces to be posited: the strong and the weak nuclear forces. The Standard Model attempts to unify these forces and it does so by describing the forces in terms of particles.

Light and the electromagnetic force are both manifestations of photons. The electromagnetic force had to be called into existence, in the same way as Newton once called gravity into existence, to explain why the negatively charged electron is drawn to the positively charged proton in the nucleus. It is this force that gives solidity to matter. It is often remarked that the atom is nearly all empty space, but it might be truer to say that the atom is filled with force fields.

The quantum field is not a field of arrows pointing the way, as a field was for Faraday; it has become in some ways more straightforward. Faraday invented a way of describing as particles in fields the behaviour of what were seen at the time as two fundamental forces of nature: electricity and magnetism. Quantum field theory is also a description of fundamental forces as particles in fields, except that now we have a different idea of what the fundamental forces are and about how we describe particles and fields. In quantum field theory everything has been reduced to particles.[3] The particle description can be said to have won out over the wave description. Even the fields themselves are made of particles: clouds of so-called virtual particles. They are called virtual because they don't appear in the input or output of the highly mathematical quantum field, but are required in order to make the explanation work. Whether or not virtual particles exist (there is no firm line that separates virtual particles from 'real' particles) in the physical world is less important to the pragmatic scientist than the fact that quantum field descriptions, collected together as the Standard Model, do a job of work. And yet if

---

[3] Faraday had wondered if atoms could be seen as concentrations in the lines of a force field rather than as physical objects, an idea that still seems revolutionary today.

we are to concede as investigators of the material world that we do not occupy a privileged position in the universe, then we are forced to accept that what we call stable or observable matter only exists because the vacuum is a soup of non-observable particles that spring into and out of existence for no reason. If we don't accept this, then we find ourselves in the curious position of saying that the kind of matter we think exists only exists because some other sort of stuff does not exist or only has mathematical significance. There is, in any case, very good evidence in support of the overall existence of these particles. Clouds of virtual particles exert a minute pressure called the Casimir effect that can be measured in the observable world, a tiny hand reaching across the divide that separates the visible world from that more encompassing world from which our local visible universe has emerged. It is as if these virtual particles exist beyond the borders of the visible world, by which is meant a patch of radiation that evolved. The universe seems to be a machine for processing information made out of some 10^{80} visible particles.

In quantum field theory, nature has been reduced to energetic fields made out of dimensionless (and, seen from the perspective of the classical world, non-existent) particles that causelessly and randomly come into and out of existence. It is pretty much impossible for the non-mathematician to understand how such a description might relate to the physical world. Quantum field theory is so abstract and mathematical that we really have little choice but to accept that such a description works and that it works because these peculiar sorts of field describe how the world really is.

Quantum field theory, with its new idea of a field, describes the electromagnetic force as a force acting in a field made from a cloud of virtual photons. The force is described as the exchange of virtual photons between the electron and the protons in the nucleus. What such an exchange of particles might look like it is impossible to say. The 'exchange' is an analogy that stands in place of a lifetime spent getting to grips with the underlying mathematics. The quantum field description of the electromagnetic force is called quantum electro-dynamics, or QED, and is known as the jewel of physics. The theory reached a high point of refinement in the 1940s after contributions from Richard Feynman, the British-born American theoretical physicists Freeman Dyson (b.1923) and Julian Schwinger (1918–1994), and the Japanese physicist Sin-Itiro Tomonaga (1906–1979). Feynman, Schwinger and Tomonaga went on to receive the Nobel Prize for their work. Initially, the theory seemed to be at odds with experimental evidence being collected at the time. But Feynman was convinced the theory had to be correct and the experimental results faulty. '[The theory] had elegance and beauty,' Feynman said. 'The goddam thing was gleaming.' This deepest interaction of light at its most tenuous accounts for much of what characterises the visible world: the existence and stability of atoms, molecules and solids.

In a similar way, a different quantum field theory was developed to explain the strong nuclear force that binds protons and neutrons together in the nucleus. Again, calling the unknown force the strong nuclear force is not an explanation, but it begins the process of explaining. It is observed that protons and neutrons are held together in the nucleus. None of the other forces of nature could be the force that holds them together, so there must be some other force of nature that does that job; and it is that force which we have named the strong nuclear force. What is peculiar about this force is that unlike the electromagnetic force and the gravitational force, both of which have unlimited range, the strong nuclear force must be entirely confined within the nucleus. The strong nuclear force is also described as the exchange of virtual particles in a field, but this time the exchange is of particles called gluons. This exchange accounts for how the quarks are bound together as protons and neutrons. This field theory is called quantum chromodynamics (QCD). The arbitrary symmetry of quark colour is what guarantees the existence of the strong nuclear force. We are willing to believe that such a symmetry exists because out of it we are able to derive a powerful mathematical description that unifies much of nature. Similarly, it is the arbitrary label of charge that guarantees the existence of the electrodynamic force that is described by QED. These two quantum field descriptions are two of the main reasons that matter exists.

The strong nuclear force finally explains alpha decay, the phenomenon that was accidentally discovered by Becquerel in the 1890s. Radioactive elements decay into other elements in various ways. Uranium, for example, naturally decays into thorium. It happens because the energetic content of the nucleus changes spontaneously due to the Heisenberg Uncertainty Principle, in a process called quantum tunnelling. Because energy can be borrowed from the vacuum, contrary to the laws of classical physics, it is possible for a subatomic particle to appear outside the nucleus. This is what happens in alpha decay. In a classical world its energy would have kept the particle confined in the nucleus forever.

Alpha radiation is a stream of high-energy alpha particles. Alpha particles are identical to the nucleus of a helium atom: two protons and two neutrons bound together. We say that an atom of, say, uranium decays when an amount of energy equal to an alpha particle is lost from the nucleus of the uranium atom and finds itself outside the nucleus where it has escaped the power of the strong nuclear force. Out of the decay of many such atoms, rays of alpha particles stream through space as alpha radiation. Strictly speaking the alpha particle isn't contained in the nucleus of the atom, the reassignment of energy is best explained by resorting to an intermediary stage involving the exchange of virtual particles, like the gluon. It is for this reason that these particles are named virtual: they largely serve a mathematical purpose to do with balancing the energy books.

The weak nuclear force, like the strong nuclear force, is confined inside the nucleus of the atom. It is this force that is required to explain beta decay, the other type of radioactive decay discovered by accident by Becquerel in the 1890s. In quantum field theory particles inside the nucleus change from one sort of particle into other sorts of particles – they are said to change flavour – and emit energy as radiation (described as yet other kinds of particles). This is what we mean by radioactive decay. The weak force is mediated by virtual particles from the zoo called W^+, W^- and Z particles. The element strontium-90 decays naturally by producing beta radiation. In beta decay a neutron spontaneously turns into a proton changing the element into another type of element. At the quark level a neutron becomes a proton if one of the three quarks that make up a neutron changes flavour. In quantum field theory it is said that a W boson (a virtual particle) is emitted by the quark that decays into a high-energy electron. Beta radiation is simply a stream of high-energy electrons.[4]

Radioactive decay is the secret to the transmutation of the elements, the secret of the philosopher's stone. It was Rutherford in 1919 who first artificially transmuted one element into another when he used a nuclear reaction to turn nitrogen into oxygen.

In particle accelerators particles are not seen directly: energetic particles decay into other particles and these particles into yet others, drawing a trail of decay that gives each particle a unique signature. Ever since Brownian motion was seen as indirect evidence of the physical existence of atoms the physical presence of ever smaller objects has become more and more tenuous.

Since the 1970s the descriptions of the weak nuclear force and the electromagnetic force have been combined into a single description called the electroweak force. It requires a symmetry conferred by a particle named the Higgs boson. A boson is the collective name for all force-carrying particles (so far, we have come across the gluon, photon, W and Z bosons). Physicists were aware that if the W and Z particles had no mass they would be indistinguishable from the massless photon. The Higgs boson was posited in order to come

---

[4] Cathode rays are also streams of electrons, but with less energy.

up with a mathematical description to explain the suspected deeper symmetry that connects W, Z and photon particles. In the electroweak theory all these particles start out as the same thing, and only become these different particles when the symmetry of the Higgs field is broken. The Higgs field slows some of the particles down and confers mass on them. This is highly significant. The Higgs boson, and the field it generates, is what changes what would otherwise be a world entirely made out of radiation into a world that has things in it that have mass. Finally, we are getting close to understanding how light can become stuff. For obvious reasons, the Higgs boson has been given the nickname the 'God particle', though in fact it is properly named for the Scottish physicist Peter Higgs (b.1929), who proposed this broken symmetry in the 1960s. For now, however, the Higgs boson has not been detected. Again, it is thought that some light might be thrown on the matter by the Large Hadron Collider.

The Standard Model hints at a unification of the electroweak and strong nuclear forces, that they converge into a single more symmetric description at very high energies. Unfortunately this energy is so high that it is well outside current experimental reach, and may always be so. But for the first time in 2,500 years, since the Presocratics first began such a search, a single unified description of reality is at least indicated. The greatest weakness of the Standard Model is that it has failed to provide a satisfactory quantum field description of gravity. Nor can it predict any of the masses of any of the particles it describes. It is wrong by 16 orders of magnitude; that is, it predicts that the particles are 10^{16} times smaller than they are actually measured, and nobody knows why.

What quantum physics actually means remains a vexed question. There are many different types of response from scientists. There are those for whom all that matters is that the theory makes accurate predictions that can be accurately confirmed, who do not care what the theory actually means. There are those who say that the problem is trivial, and those who claim that it goes to the heart of who and what we are.

The central questions are *where* (at what sizes) and *how* does the quantum world become the classical world? In 1927 Schrödinger encapsulated the problem in his famous thought experiment known as Schrödinger's cat paradox. By placing a cat and some radioactive material in the same box, Schrödinger's intention is to bring together an obvious classical object (a cat) and an obvious quantum object (something radioactive). This *thought* experiment (no animals were harmed during the making of this experiment) is contrived in such a way that if the radioactive material decays then a phial of poison is broken and the cat dies. According to the traditional[5] interpretation – the so-called Copenhagen interpretation – of quantum physics, it is impossible to say if the radioactive material has decayed until a measurement is made. In the Copenhagen interpretation, when a measurement is made the wave function collapses. That is to say, the system takes on a particular value (the number attached to the measurement) out of an array of possible values with various probabilities attached to the different possible outcomes. As Heisenberg pointed out, quantum objects don't have a history until a measurement is made. There is no history in the quantum world. At quantum sizes, Heisenberg tells us, even single particles are unpredictable.

So who gets to do the measuring? Schrödinger's point is that we say that *we* do the measuring when we open the box and take a look to see if the radioactive material has decayed or not. If it has decayed, the history of its decay is only revealed at that moment. But if this is true, what state was the cat in

---

[5] The first significant attempt to try and explain what sort of physical reality quantum physics represents was thrashed out by Niels Bohr and Werner Heisenberg in Copenhagen in 1927. In the same year, the triennial Solvay Conference held in Brussels, named for the Belgian industrialist Ernest Solvay, was also devoted to the interpretation of quantum physics. Of the 23 delegates, 17 were either Nobel laureates or would become Nobel laureates, Marie Curie unique among them as a double Nobel laureate. It was at this conference that Einstein declared in reference to Heisenberg's Uncertainty Principle that 'God does not play dice', and to which Niels Bohr replied: 'Albert, stop telling God what to do.'

in the meantime, before the observation was made? It is all very well for an object in the quantum world to live in this partial state, but what does it mean for a classical object to remain in an indeterminate state? Quantum physics appears to marry the observer to the observed.

It has since been shown that it is possible to delay the decay of a radioactive atom by looking at it, in what seems to be a quantum proof of the old adage about the watched pot. Constant observation stops the wave function that describes the quantum object from evolving and so delays the possibility of the atom decaying. But do we really get to wield such power over reality? Does the world only mean something when humans make observations? And if humans, why not other creatures with consciousness? Cats even. The idea that human consciousness is so privileged is about as anti-Copernican a notion as could be conceived. Mankind might no longer be at the centre of the universe, but here is something worse: mankind gets to determine the nature of reality.

It turns out that we do wield such power over quantum objects, but it is very difficult to isolate a quantum object from the rest of nature, and it is this fact that offers a way out of this dilemma. It is currently possible to isolate molecules as large as buckminsterfullerene molecules (made of 60 carbon atoms). In laboratory conditions it has been shown that such a molecule can be made to pass at the same time through two slits made in a screen, in a set-up largely unchanged from Young's double-slit experiment. In other words, scientists have proved that they can reveal the quantum nature of a molecule as large as buckminsterfullerene. A few decades ago such a possibility would seem to be beyond any conception a materialist might have had of the world. But we now know that such magic is possible. Scientists have found a way of isolating the buckminsterfullerene molecule from the visible world and preserving its quantum nature. We could say that it exists as some sort of wave of probability and it is this wave that passes through both slits. Only when we make a measurement and ask where the molecule is does it become a visible object in the visible world with properties of thingness. It appears to us that the only way the molecule could be where it turns out to be is if it had passed through both slits. But it would be a mistake to imagine that the molecule split into two existences as it passed through the slit: that would be to elevate the idea of what we mean by existence in the visible world. Observed reality is a local idea of existence. There is a deeper understanding of existence, and that is existence in the quantum world, which we can only measure statistically and partially in the observed world. Quantum reality is not about being in two or more places at once, it is about not being anywhere that we call place. Such existence is not non-existence, it is an extension of what we mean by existence.

The separate natures of these two worlds are apparent when we isolate objects from the world of classical existence. However, to keep a quantum object separate from nature takes a lot of effort. Even an object made of just 60 atoms requires extreme cold to prevent the object becoming something classical. The passing from quantum reality to classical reality is called decoherence. Recently we have escaped the anti-Copernican nightmare of privilege inherent in Schrödinger's cat paradox by understanding that it is Nature that does the observing, not humans. The wind and sunlight measure the tree, and convince us that the tree exists even when we are not there to observe it. It is the interaction of quantum objects with the environment that produces what we understand as classical objects, such as cats and tables. We have found out ways of preventing quantum objects from decohering, but it is a very hard thing to do. There is so much of nature, and it is determined to embrace, to measure, everything. We might keep a single molecule from decohering, but there seems to be little chance that we could isolate a cat made up of 10^{27} atoms.

Information leaks from the one reality into the other. Decoherence seems to be how nature spreads information around the universe, and may even be connected to why we perceive time flowing forward. The flow may be the flow of that river of information that prevents quantum objects from being isolated from nature for very long.

In the 1920s when the interpretation of quantum physics was first up for grabs, the nature of existence linked physics to Eastern mysticism, a relationship that has troubled some scientists ever since. Certain Eastern modes of thought tell us that the world is not made up of things or nothing, but it is a web of interconnected phenomena. Quantum physics has arrived at the same place. When we separate out objects, it is the separateness of them that is the illusion. The deeper reality is the inseparable world from which they have been extracted. The scientific method puts rings round phenomena in order to describe them, and then out of those descriptions finds out the laws of nature that encompass and connect ever more phenomena. In recent decades it has become apparent that science might be capable of finding out a single description of nature, a description that shows how an undivided web of phenomena appears to be manifest as separateness. Because the methodology starts out with an idea of separateness, it is easy to believe that the separateness is the deeper reality, and forget that the deeper purpose is to understand the inseparability. Quantum physics has brought us to a place where we can begin to understand what certain mystics have always understood: how a world that appears to be made of separate things arises out of a world of inseparability (of no things).

There are several competing interpretations of quantum physics. For example, the many worlds interpretation was put forward by the American physicist Hugh Everett (1930–1982) in 1957. Rather than have the wave function collapse only when it is measured, he hypothesised that every possible quantum event happens. All possible worlds exist alongside each other in which every possible outcome is realised. In the many worlds scenario reality is nowhere. It is an edge between all possible quantum worlds. The radioactive material and the cat move off into many separate universes, in some of which the cat is alive and in some not.

The physicist David Deutsch (b.1953) is a strong supporter

of the theory. He argues that there may soon be evidence that these many worlds do actually exist. Quantum computers are thought to be only a decade or so away. Instead of the binary 0 or 1 system on which classical computing is based, quantum computing works on the principle that quantum objects can be held in a state of many possible outcomes all at the same time, and so quantum computers can perform many operations at the same time. In fact a quantum computer can, in theory, crack a modern security system in seconds, a system that could not be cracked by classical computers if every particle in the visible universe were to be turned into a computer working on the task for the whole life of the universe. If quantum computing becomes a reality, then where, David Deutsch asks, are we to suppose that the quantum computing takes place? A quantum computer can only be drawing on the computing power of parallel universes.

On the down side, the many worlds hypothesis is about as extravagant as it is possible to be. Nowhere is the methodology of science written down. It had evolved. But it is generally agreed that the best scientific descriptions are also the most parsimonious, that is, are constructed using the least number of principles. This belief is known as Occam's razor, after the English Franciscan friar and philosopher William of Occam (c.1288–c.1347) who first expressed the principle. The many worlds theory predicts the existence of a very large, perhaps an infinite, number of parallel universes. Such a prediction is enough to convince some scientists that the theory is worthless.

Although Einstein was one of the architects of quantum mechanics, he could not believe in a world in which electrons do not exist in the absence of observers. He believed that there was a deeper reality beyond quantum physics that would retrieve the idea of completeness, just as the gas laws, though statistical, had retrieved the possibility of complete determinism. In theory, it ought to be possible to describe gas using Newton's laws of motion. In practice, there literally isn't enough time in the universe to fully describe the motion of all the molecules even of a single jar of gas. But by treating the molecules statistically (rather as we might a crowd of individuals), the behaviour of the gas as a whole can be understood without knowing the behaviour of individual atoms. We preserve the possibility of a complete determinism but make do in practice with a statistical description. Einstein thought that the quantum world would ultimately be revealed as complete in this way.[6]

Einstein was particularly intrigued by the work of the American theoretician David Bohm (1917–1992). In the classical world, wholeness is the integration of our understanding of separate things. David Bohm turned this on its head and said

---

[6] At the end of the twentieth century, chaos theory undermined the determinism of classical mechanics, just as earlier the gas laws had done. Chaos theory shows us that there are systems in nature so finely tuned, as we are beginning to understand to our cost, that the tiniest differences between two systems that in all other respects are identical lead to very different outcomes. Chaos theory shows that many natural systems are described by unstable mathematical equations that are only deterministic in theory. The weather is such a system. No matter how minutely we think we can describe the weather at a particular moment, it can unfold into dramatically different weather from that which we predict, because the smallest imprecision in one of the measured variables may lead to a very different outcome. The effect is known as the 'butterfly effect': by not taking account of the butterfly's disturbance of the air we may be unable to account for the subsequent hurricane. The French philosopher Blaise Pascal (1623–1662) expressed a similar idea when he wondered how differently history might have run had Cleopatra's nose been a different length, and the English historian A. J. P. Taylor (1906–1990), when he reflected that maybe if Archduke Ferdinand's carriage had not accidentally turned down the street where he was then murdered, the First World War might not have arisen. For the sake of a nail ... the battle was lost. Chaotic equations are deterministic; it's just that they cause us to rethink what we mean by determinism. It happens again in quantum physics. Heisenberg tells us that even single atoms are unpredictable. Nevertheless, quantum systems are described by wave functions that are deterministic, even though the reality that is described is itself unpredictable. It is often said that free will is impossible in a deterministic world, but it's as if the world has set itself up in such a way that the *illusion* of free will is guaranteed. It's the complexity of the evolved world that makes that *illusion* compelling, as if nature is determined to save us from existential despair.

that it is actually wholeness that determines the behaviour of things that we see as separate things. It is as if we see the world from separate camera angles and take such perspectives as evidence of different phenomena, when in fact the camera angles are, if we but knew it, different perspectives on the one reality.

Einstein, along with the Russian physicist Boris Podolsky (1896–1966) and the Israeli physicist Nathan Rosen (1909–1995), devised a thought experiment called the Einstein–Podolsky–Rosen (EPR) paradox to show that the reality described by quantum physics could not be incomplete as it is in, say, the Copenhagen interpretation. Without going into details, EPR asks us to imagine a set-up of just two particles which have been produced in such a way that they are guaranteed to have some equal and opposite property, say spin. According to the Copenhagen interpretation of quantum mechanics, when we decide to make a measurement (in this example, of spin) the wave function of the system collapses into one of various possible outcomes. The paradox arises out of the fact that when we measure the spin of the one particle we are guaranteed from the way the system was set up to know what the spin of the other particle is, and this would be true even if the particles are millions of miles apart when we make the measurement. Such a possibility would seem to break Einstein's own law in which there can be no instantaneous transference of information because of the limit set by the speed of light. Einstein called such a possibility 'spooky action at a distance'. How could the other particle 'know' that the wave function had collapsed? In a way, the paradox is not so different from Schrödinger's cat paradox in that its primary purpose is to highlight some sort of irreconcilable discontinuity between the macroscopic world and the microscopic world as described by quantum physics.

The French physicist Alain Aspect (b.1947) found a way to turn the EPR thought experiment into a real experiment. Unfortunately for Einstein (and Podolsky and Rosen), in 1982, after years of careful measurement, Aspect proved that the quantum world *is* incomplete. And more recently, sophisticated experiments by Nicolas Gisin at the University of Geneva, using a fibre-optic network that runs miles around Lake Geneva and into nearby towns, have also showed that instantaneous communication is possible among quantum objects. Einstein was wrong, and the world really does exist imprecisely.

The scientific method has found out ingenious ways of extending what we mean by existence and reality, even beyond what we call the visible universe. Ultimately, it is ever more powerful mathematics that leads the way, to which is added the imperative that a material interpretation of that mathematics is essential if the scientific method is to reach further still. The outcome, however, is a kind of materialism that is so bizarre – the belief in an infinite number of parallel worlds, for example – that the gap between mystery and mystical barely seems apparent. What divides now the mystic and the materialist?

Something and Nothing

On Margate sands.
I can connect
Nothing with nothing.

T.S. Eliot, 'The Waste Land'

The most famous equation in science is the equation $E = mc^2$ that describes the equivalence between energy and mass as predicted by Einstein's special theory of relativity. Energy (E) and mass (m) are the same thing, the equation tells us. And more than that, there is an unchanging number found in nature that shows us precisely how much energy there is in any given amount of matter. That unchanging number, c, is the speed of light, the number of metres that light covers (in a vacuum) per second, which is roughly 299,792,459. In Einstein's equation we see that the large value of the speed of light is squared, making it very large indeed: roughly 8.99×10^{16}. Here, then, is a clue as to why even a small amount of mass is equivalent to a lot of energy: the secret of the atom bomb. Einstein originally wrote the equation as $m = E/c^2$ as if mass were the deeper principle. $E = mc^2$ is the same equation, but writing it out in this form is an aesthetic choice that makes it clear that it is energy that is in need of the deeper explanation. We know that mass is the stuff that distorts space-time, and it may also be what is conferred on energy by the Higgs field. We are less clear what energy is. We know that it comes in many forms and we know how those forms change one into another, but ultimately we don't know what energy is. Sunlight, for example, is turned by the process of photosynthesis into plant matter, some of which, over vast periods of time, has been turned into coal. When we burn coal, the chemical energy in the bonds between the molecules that make up the coal is turned once more into light and heat. On his travels Gulliver met scientists who were trying to retrieve sunlight from cucumbers. They were not so crazy. If only they had had enough time and large numbers of cucumbers.

Sometime in the 1940s, Einstein was walking through Princeton in conversation with the Russian-born theoretical physicist and cosmologist George Gamow (1904–1968), who happened to mention that he had come to the realisation, while pondering Einstein's discovery that energy and matter are equivalent, that a star could be created out of nothing at all since the energy of its mass is exactly balanced by the energy of its gravitational field. Einstein was so taken aback by this insight that, as Gamow reported, 'since we were crossing a street, several cars had to stop to avoid running us down'. If the universe, as large as it is, is merely a hierarchy of stars, then it too could have emerged out of nothing. Its overall energy is zero. Parmenides and King Lear were wrong: the universe is something that comes from nothing.

We find ourselves in a curious reversal. The universe seen as arrangements of stars turns out to be nothing, yet the smallest amounts of space are roiling with energy. We have two quixotic descriptions of nature: relativity, from investigation of the universe at large, and quantum physics, from scrutiny of the universe at its smallest sizes. Taken together they tease us with their hints of both unification and opposition.

*

The 'cobbled together' Standard Model of quantum physics accounts for three of the four forces of nature – the electromagnetic force and weak nuclear force combined into a single electroweak description, together with the QCD description of the strong nuclear force – and hints at a possible unification of these three forces at very high energies if another layer of particles, called supersymmetric particles, is added to the particle zoo. Supersymmetric particles allow scientists to find a connection between those particles used to describe the forces of nature (bosons) and the fundamental particles that describe matter (fermions). Unification of the electroweak and the strong nuclear forces comes at a high price: every single particle in the particle zoo, and there are hundreds of them, needs to have a supersymmetrical partner. For matter particles such as the quark and electron the supersymmetrical partners are called a squark and a selectron, and for force-carrying particles such as the photon and gluon they are called the photino and gluino. The major failing of this description is that none of these hundreds of particles has been detected, but at least a mathematical unification is possible, which is a beginning. Once more, it is hoped that some evidence will emerge one way or the other from the LHC. Some scientists are eager to find evidence that supersymmetric particles do not in fact exist. There are those who feel that the Standard Model is so out of control that surprising and unpredicted physical evidence would be more useful at this point, evidence that might point towards some new way of describing reality.

Supersymmetry also shows how the three forces described by the Standard Model might be unified with gravity. The Standard Model can be extended to include a quantum description of gravity by predicting a force-carrying particle called the graviton and a supersymmetric particle the gravitino. Neither particle has been detected, and in the Standard Model the existence of the graviton produces some troubling infinities in the mathematics, usually an indication that the theory has broken down. Though the graviton has not been found, at least we know what it must look like: we know what kind of spin it must possess and we know that it must be massless like the photon. We also know that it must be very difficult to detect because of gravity's puzzling weakness. Gravity is 10^{40} times weaker than the force of light, and so the predicted gravitons will hardly interact with matter at all. It has been argued that it would take a device the size of Jupiter with lead shielding several light years in thickness to detect them, and that even then we probably would not be able to detect enough particles for the evidence to be convincing. Though supersymmetry lacks physical evidence, the mathematics predicts a unification of all the forces of nature at very high energies: at around 10^{19} electronvolts.[1] This is a muted cause for celebration since even the LHC will reach energies of less than 10^{13} electronvolts. Further, even modest increases in energy come at huge cost, putting this unification energy seemingly forever beyond reach.

One of the most hopeful and frustrating attempts to unify gravity and particle physics is string theory. String theory requires a plunge downwards to a world made out of strings of vibrating energy 10^{-35} metres across, 17 orders of size smaller than the maximum size of the electron and quark. A string is as small compared to an electron as a mouse is compared to the solar system.

Seen from the perspective of strings, the particles of the particle zoo only look like points because we are seeing them from a distance. Closer to and at higher energies, what we had taken to be fields and particles reveal themselves to be made out of vibrating lengths of pure energy. String theory attempts to smooth out the violent nature of the quantum world in order to find some deeper symmetry hidden behind its apparent randomness.

By turning quantum theory into a smooth theory, string theory tries to marry a description of the smallest things to

---

[1] A single electronvolt is the charge on the electron multiplied by a single volt. It can also be defined as the amount of energy needed to accelerate an electron through a potential difference across a conductor of a single volt.

the smooth description of the larger world that is general relativity. Theories that hope to make such a unification are called TOEs (theories of everything). It isn't yet clear if string theory is such a theory or not.

It will come as no surprise to learn that such attempts at unification come at a price. String theory began life as a theory in 26 dimensions, and was reduced to a theory of 10 dimensions called superstring theory after the discovery of a symmetrical mathematical object called a Calabi–Yau shape, which was named after the Italian-American mathematician Eugenio Calabi (b.1923) and the Chinese mathematician Shing-Tung Yau (b.1949). Unfortunately, the 10-dimensional theory came in five different forms until, in 1995, the five theories were united into a single description in 11 dimensions called M-theory, which is what we mean when we talk of string theory today.

String theory is so complex that there are around 10^{500} possible universes that it might describe. The theory provides no obvious means of showing how to choose which of those solutions describes this universe. We also have to ask why, if reality exists in 11 dimensions, we are only aware of four of them, three of space and one of time. If the additional dimensions are space dimensions, it is thought that the extra seven dimensions that we are not aware of are rolled up so tightly that they cannot easily be seen. A wire cable viewed from a distance looks like a one-dimensional line, but an insect walking on the wire knows differently: it can walk around the wire, seeming to disappear and reappear from the perspective of a distant observer. The disappearance of nature's other seven dimensions is an extension of that illusion but in more dimensions. If the extra dimensions of string theory include *several* time dimensions the description becomes even more bewildering.

The general idea is that the quantum world may only appear to be chaotic and violent because a smoothly changing world of 11 dimensions looks like that to us in our illusory world of four dimensions. The world becomes more symmetrical when we see that it is constructed in 11 dimensions.

It is possible that evidence for the existence of these additional dimensions might be found using the LHC. (The LHC is going to be very busy.) It is conjectured that certain so-far undiscovered particles, which would ordinarily be revealed as spikes of energy, might instead reveal themselves as 'patterns of missing energy', because they are hidden from view in tightly curled up dimensions of space. It has to be said that this is a strange sort of test. It requires some nerve to believe that the arcane mathematical descriptions of string theory not only have physical meaning but that that meaning can be confirmed by the non-appearance of some predicted but hitherto unobserved particle.

Both the Standard Model and string theory require supersymmetry, and as we have seen there is no evidence that supersymmetric particles exist; but at least in string theory the infinities that dogged the description of gravity in the Standard Model can be dealt with. In fact the triumph of string theory is that certain strings have exactly the right properties to be gravitons. The theory actually predicts the massless graviton and its required spin. Whether or not it is a theory of everything, string theory is a quantum theory of gravity.

In general relativity, we already have a description of gravity, so if we are to have another description of gravity in quantum physics then somehow the two descriptions must marry together as the same description: general relativity must be married to quantum physics. Scientists are beginning to understand at what point the quantum world becomes the world of large things: it is very hard to isolate even a few molecules as quantum objects. Symmetrically, we might ask: at what point does the description of large things become a quantum description?

Because the Big Bang solution of general relativity is a description not only of the universe at large but of a universe that grows, here is a way to find out where our two descriptions of nature might meet up. There is in fact a surprisingly straightforward way to find out what quanta of space and time, if they exist, must look like. If we run the Big Bang expansion backwards in time, the universe seems to have emerged out of a dot of infinitely dense energy or matter, which is just another way of saying that the Big Bang theory breaks down at the moment of creation: a veil descends and obscures the origins of the universe. Fortunately, because we also have a theory that describes the world at its smallest sizes, quantum theory can tell us where general relativity breaks down. By shoving quantum physics and the Big Bang theory together, we can find out how small the universe can be before it becomes a purely quantum object.

*

Though our energetic and material world is in perpetual flux, mysterious numbers crop up in the laws of nature that mankind has devised or found out to describe that activity. Numbers like c, the constant that denotes the speed of light, G, the gravitational constant that measures the strength of the gravitational force, e, the charge on the electron, and h, Planck's constant that determines the size of a quantum of energy, are seemingly unrelated to each other, and crop up in different laws of nature. Some scientists believe that if we are ever to derive a unified theory of everything we must ultimately understand what connects these numbers each to the other. The constants of nature are ugly numbers expressed in ugly collections of symbols: G, for example, is 6.67259×10^{11} m^{3}s^{-2}kg^{-1}. But a simple manipulation of just a few of these constants allows us to find natural units of measurement called Planck measures (not to be confused with the Planck constant itself): the smallest units of measurement that could mean anything.

Planck length is 4.13×10^{-35} metres, the shortest length possible that has any meaning as length. Since we know what speed light travels at, and that light is the fastest means of communication in the universe, we can also calculate Planck time: the shortest increment of time that has any meaning. It is simply the length of time it takes light to travel across a unit of Planck length; a straightforward calculation that tells us that Planck time is 1.38×10^{-43} seconds. In some sense, then, a universe with time in it must have already been 10^{-43} seconds old when it was created, since time had no meaning before that. And a universe of any size must have been already 10^{-35} metres across, since a universe smaller than that has no meaning.

Quantum physics tells us that it is more useful to talk of the energy of an elementary particle rather than its size. We know that particles can reveal extent in space if we insist on that knowledge, but now we also know where size itself disappears altogether from the world.

String theory is woven out of lengths of energy that exist right at the edge of the meaning of length. The vibratory quality of these tiny threads of energy is what replaces particles and fields in the quantum field description of nature. String theory resonates with ideas attributed to the Pythagoreans: the discovery that there are simple mathematical relationships between the sounds made by plucked strings of different lengths. In the high mathematics of string theory the strength of the vibration is what we see in the world as mass and the patterns of vibrations are the fundamental forces. It is as hard to say why this should be so as it is to say why a cloud of virtual particles describes the fundamental forces. The validation is in the mathematics, and we poor non-mathematicians can only take such descriptions on trust.

String theory is not the only game in town, and some physicists think it has been a disaster for science, taking many of the best brains off on a wild goose chase. The theory of loop quantum gravity is another attempt at a quantum theory of gravity. In string theory the fabric of reality is woven out of strings of energy. In quantum loop gravity the material is woven from quanta of space and time; a way of knitting the space-time of the relativity theories.[2] It is a guess on the part of theorists that there are actual quanta of space and time,

---

[2] In quantum mechanics space and time are rigid as they are in Newtonian mechanics, not malleable as they are in the special and general theories of Relativity.

but since everything else in quantum physics is chopped up into little pieces – spin, charge, colour, mass, energy, and so on – then why not space and time too? In quantum loop gravity the particles of the particle zoo are shapes braided together out of Planck lengths of space. Stephen Hawking has taken these ideas even further and suggested that the history of the universe itself is quantised. How we measure the universe determines what history it turns out to have had. When we measure, we change our past. Hawking and the American physicist James Hartle have together constructed a theory that turns time into space at very high energies, another way of dealing with the problems of how we talk about the beginning of the universe. Hawking and Hartle argue that it is meaningless to ask the question how the universe began since there was no such thing as time then. 'In the beginning' is not the beginning of this story because time becomes a space dimension at high energies. But this is to stray into worlds of such great abstraction, even of imaginary time, that we might begin to risk madness.

What quantum theories of gravity do is tell a story of how the universe began, even if it is to say that it did not begin in time. The energy required to see the quantum nature of gravity is enormous and quantum physics tells us how enormous. In theory, if we could put this energy into the vacuum, we could retrieve the conditions that existed at the beginning of the universe. By putting increasing amounts of energy into the vacuum, we approach the original quantum nature of the universe. This is what we see with more and more powerful particle accelerators: the conditions that pertained closer and closer to the beginning of the universe.

Running the Big Bang backwards shows us that all the matter in the universe becomes more and more energised until it becomes pure radiation. Current theories seem to suggest that when the universe was a pure quantum object its symmetry held all the forces of nature together as a single entity. In the words of the Belgian-born American physicist Armand Delsemme (b.1918), the universe arose out of 'the spontaneous rupture of the pre-existing grand symmetry of nothingness'. The immense energy and symmetry of the vacuum broke, and our visible universe was the result.

The ancients used the power of poetry to communicate knowledge. Our modern-day materialist creation story has become so abstruse that it may need a poet to do it justice. In scientific discourse the poetry is in the mathematics, and the same language judges them alike: symmetry, elegance, simplicity, brevity, subtlety, profundity are the highest qualities of both means of apprehending reality. Mathematics has been the language of science ever since. But that language is increasingly untranslatable even among large groups of scientists. It is not even a single arcane language but many arcane languages each spoken by some small tribe of specialists.

It is not, however, the complexity and sophistication of the mathematics, nor the ability to find and name hundreds of particles, that is impressive about quantum theory, rather the opposite. The difficulty and fragmentation is the high price we are willing to pay for the suggestion of simplicity and unification underlying that fragmentation: that the fragments are the shards of some beautiful object that fell and broke into pieces.

A theory of everything appears to promise a universe that began as some form of perfect symmetry that was broken. Everything else followed. Perfect symmetry, it seems, is not a quality of this world but the condition from which it arose. This world is one of broken symmetry. The ancient Greeks knew this. They understood that the most beautiful things are not those that are perfectly symmetrical but those that are nearly so. Only in the heavens beyond is the perfect symmetry of a sphere attainable. The history of science can be seen as the pursuit of symmetry. 'Pattern recognition', says the American physicist Bob Park (b.1931), 'is the basis of all aesthetic enjoyment, whether it is music, poetry or physics.'[3]

The perfect symmetry from which the universe arose has

---

[3] Quoted in the *New Scientist*, 9 December 2006.

been likened to a sharpened pencil standing on its point, a state that is too symmetrical to last. The pencil must immediately fall but we cannot know in which direction it will fall. Surely this grand moment that wasn't a moment and didn't happen anywhere (because location was not a quality of this place that is not a place) deserves a better analogy. Whatever this symmetry was, it broke for reasons not yet understood.

In 1952, George Gamow named the period before the Big Bang the Augustinian era, after St Augustine who wrote that time was created when the universe was created. At 10^{-43} seconds a patch of symmetric radiation that is our universe began in time. For a moment the four forces of nature were held in symmetry.

10^{-43} seconds to 10^{-36} seconds

The universe expands and the temperature immediately begins to fall from the highest possible temperature, 10^{32} degrees (called Planck temperature). At some time in this 'era', symmetry breaks and gravity enters the world. From our vantage point we believe that more energy is required to unite gravity with the other forces of nature than is required to unite the other three forces to each other, and so we must also assume that it is gravity that will be the first force to break away from that symmetry. Symmetry breaking is also called phase changing. In our local world we see a phase change when water changes from its more symmetric form as a liquid to a less symmetric form as ice.

By the end of this era the temperature has fallen to 10^{27} degrees. The history of the universe is usually told in time but it could just as well be told in falling temperature. The expansion of the universe is inextricably connected to both its cooling down and its evolution. The universe expands in time and what is in the universe spreads out, cools down and evolves.

The difference between a universe that is 10^{-43} seconds old and one that is 10^{-36} seconds might seem to be neither here nor there, but measured in Planck time the difference is between a universe that is a single Planck unit of time old and one that is 10^7 (10 million) Planck units old.

Because the strong and electroweak forces are still unified, in theory the only particle in the universe at this time is the Higgs boson.

10^{-36} seconds to 10^{-12} seconds

When the universe is 10^{-36} seconds old the strong nuclear force begins to break away from its symmetry with the electroweak force. The number of particles increases to include the W and Z bosons that mediate the electroweak force.

Sometime between the universe being 10^{-36} seconds old and 10^{-32} seconds old (between 10^7 and 10^{11} Planck time units old) the universe not only expands, it inflates – a theory first put forward in 1982 by American physicist Alan Guth (b.1947). The expansion of the universe as described in the Big Bang theory is of a steady increase. But in this era the universe is conjectured to have expanded exponentially, doubling in size perhaps a hundred times in what may seem to us to be a shockingly short period of time. It is claimed that when the universe was less than 10^{-32} seconds old it doubled in size every 10^{-34} seconds.

Doubling the universe a hundred times over doesn't sound like much, but it takes the universe from something existing in a quantum landscape to something the size of a grapefruit.[4]

---

[4] The mythical story of the invention of chess, perhaps in India 1,400 years ago, shows how the repeated doubling of a number of things quickly takes us into the realm of the unfeasibly large. A peasant makes a present of his invented (or discovered) game to the emperor, who, being mightily pleased, asks the peasant to name his reward. The humble peasant asks only that he receive a portion of rice to be measured out on the squares of the chessboard, a single grain on the first square and twice as many grains on each of the subsequent 63 squares. The emperor, evidently not being of a mathematical frame of mind, readily agrees; he is even pleased he has got off so lightly. Sacks of rice are brought forth and the measuring out begins. One

I don't know why a grapefruit, but among scientists and science writers this is the fruit of choice. As there is no outside from which to judge, nor any observers inside, it is hard to say in what sense the universe is the size of a piece of fruit. The universe began as high-energy radiation (light), and in a sense it was already infinite and timeless. From an outside perspective (which doesn't exist) we would see that the clock never ticks. The universe that has grown to the size of a grapefruit can only mean anything if there will be some future observer to give the grapefruit universe meaning. The grapefruit is the visible universe, but it is also possible that the universe began infinite in size and expanded. What we call the universe is merely as far as it is possible to see in that landscape.

The theory of inflation, an ad hoc addition to the quantum theory, helps solve some troubling problems that the general theory of relativity fails to account for. The cosmological principle that Einstein added to general relativity treats the universe as if it is smoothly filled with matter, which it clearly isn't, at least not locally. We see matter clumped together in

grain, 2 grains, 4 grains, 8 grains, and so on. Thirty-two grains for the 6th square, 512 on the 10th, but 134,217,728 grains for the 28th square. The emperor is presumably by this point furious. The last square alone must contain 2^{63} grains, which we can show to be more than every rice crop in the history of the earth. If you can be bothered to make the calculation, 2^{63} is 9,223,372,036,854,775,808, or close to 10^{18} – a billion billion grains. A thousand gains of rice measures out to be about 25 grams, so there are about 40,000 grains in a kilogram. This gives us about 230×10^{12} kilograms, or 230 billion tonnes of rice. A little investigation shows that China's total rice yield for 2005 was 31.79 million tonnes and China accounts for 40 per cent of the world market, which would make the world market about 75 million tonnes. If the world produced as much rice as this every year, the last square of the board would account for about 3,000 years' worth. But rice production must have been minimal millennia ago and have grown exponentially in recent times, matching the growth of the world's population. Indeed modern short-stemmed rice plants have only existed since after the end of World War II. We can confidently assert that much less than the amount of rice the emperor agreed to place on the last square of the board has been produced since mankind began to farm about 12,000 years ago.

the structures with which we have become familiar, as stars, galaxies, and clusters of galaxies. Only at the very largest orders of size does the stuff in the visible universe appear to be smoothly distributed. General relativity cannot account for why the universe is clumpy at all lesser orders of size, but quantum theory can. The randomness and violence of the quantum world is the clue.

In the vacuum countless bubbles of energy come into and out of existence creating what is sometimes called quantum foam. It is from out of this foam that our universe emerged. Quantum field theory allows these bubbles to grow to about 10^{-27} metres before they disappear back into the vacuum. Occasionally, for reasons we do not understand, a bubble escapes the vacuum. One such bubble became our universe. In other words, what we have called the universe is merely an inflated patch of a quantum landscape where size is meaningless. The universe is possibly infinite in size and the quantum landscape it escapes from is beyond the meaning of size.

This quantum landscape is a place where many universes, perhaps an infinite number, live and die. To say that this quantum world is larger than any of the universes that emerge from it cannot mean anything. Size is something that only means something locally within our visible universe. Size may not mean anything even in other parts of our universe beyond the horizon of what we call the visible universe, let alone in other universes that escaped the vacuum.

As Copernicans determined to undermine the idea that we are at the centre of anything, this inflationary model brings with it some reassuring consequences. We must suppose that there are many other bubbles – perhaps even an infinite number – that expanded into other patches in the quantum landscape, and with different laws of nature. In this way we are assured once more that our position even in this bizarre quantum landscape is not privileged. What we have called the visible universe is clearly a local phenomenon; even what we have called the universe turns out to be a local phenomenon. The 'true' universe is a quantum landscape out of which our universe (and the patch of it we call the visible universe) and many other universes arose. Just as atoms are not indivisible even though their name suggests otherwise, so is the universe no longer the word that defines everything that exists. The multiverse is what scientists have opted to call this new-found land.

Our local universe, then, is a gift of a random and causeless fluctuation in a quantum landscape. What lies beyond the furthest reaches of the universe in both directions, at the smallest and largest sizes, lies beyond our current understanding of the laws of nature. Historically, as we deepened our description of the universe, the universe grew in size; now the universe grows beyond the idea of size itself. The multiverse, or whatever we call the next manifestation of the universe, may always be beyond our ability to describe it. In our attempt to find it out, the universe grows ever subtler.

Whether inflation is more than a brilliant mathematical trick, it is too early to say. Strictly speaking it is a model, or a hypothesis, not a theory. It is not yet clear how the mathematics can be turned into experimental tests. The inflation model presents the experimental physicist with difficult challenges. It currently has greater explanatory power than it does predictive power. The inflation model is, however, widely accepted as the best scientific description we currently have of how the universe might have got going. What is particularly encouraging is that it solves various seemingly intractable problems. Inflation translates the random violence of a quantum event into a universe of clumpy matter. The clumpiness inherent in the quantum foam becomes the clumpiness of a galaxy or a cluster of galaxies. A quantum pattern is repeated in the almost instantaneous action of inflation through all orders of size in the visible universe. Inflation is permitted to happen faster than the speed of light because no information is transferred. The speed of light puts a limit on how fast information can be transferred across the universe, but if we imagine that the universe inflated, whatever is in it effectively stays put. The universe simply gets blown up into larger scales of size. Inflation also explains why nature looks similar at all sizes.

Fractal geometry is the branch of mathematics that classifies objects that have the strange property whereby their parts look like the whole. In nature a cauliflower is fractal, as are mountain ranges, snowflakes, clouds and ferns. There are also resonances between fractal objects: a coastline looks like the edge of a leaf, and a cyclone looks like a spiral galaxy. Inflation ensures that the universe is fractal at all sizes below the largest sizes. The cosmological principle tells us that at the largest sizes the universe is smooth. When we look at the universe as a whole, it is as if it has the texture of a piece of bread (I mean highly processed white bread), but at all smaller sizes the bread has structures made out of crumbs. (You see, I can't resist a homey analogy either.) The same quantum pattern inflates and becomes the universe at different orders of size. The appearance of the universe as it is now is explained by the randomness and causelessness of the quantum world.

The inflation conjecture does something else: it explains why space appears to be incredibly flat. By 'flat' is meant that the angles of a triangle add up to 180°, as they do when we do geometry on a flat piece of paper. The angles of a triangle drawn on a ball do not add up to 180° and a ball is not flat. If the cosmological principle tells us that the universe at its largest dimensions has the texture of a slice of bread, common sense might tell us that space is unlikely to be flat like a slice of bread (or a piece of paper) because the universe is full of mass and we know that mass distorts space-time. In the early days of cosmology after general relativity was first posited, it was thought that the mass in the universe had bent space back on itself. In such a 'closed' universe it would be possible to set out on a journey and eventually – if we travel for enough – find ourselves back where we had started (as if we had circumnavigated the globe, but in three dimensions of curved space rather than the two dimensions of the curved surface of the earth). At that time, too, it was thought that, ultimately, the mass in the universe would overcome the expansion of the Big Bang and draw the universe back together to reapproach the quantum state from which it had arisen. But since the 1960s, observations tell us that the most likely outcome is that the universe will expand forever, and that space is not only flat but it is incredibly flat.

Inflation offers a simple, perhaps too simple, explanation. Inflation smoothes out space as if a wrinkled balloon is being blown up. The flatness of space is established before the presence of mass has a chance to influence the geometry of space.

If the mathematics of the quantum field description of inflation is truly a description of how the world really is, and not just a mathematical model, then there must be yet another unobserved particle hidden in the vacuum. What else, but the inflaton? The Higgs boson brings mass into the world, the charge on the electron guarantees the electromagnetic force (described as a field of virtual photons), colour guarantees the strong force (described as a field of gluons), and so the inflaton, if it exists, will describe the force field that inflated the universe.

Whereas the Big Bang theory predicts that all the matter that now exists was already there at the start, infinitely compressed, the inflation theory allows the universe to be created from about 10 kilograms (or a little over 20 pounds) of matter, opening up the tantalising or terrifying prospect that it could be possible to create a universe in the laboratory. In the inflation model all the matter in the universe is created out of the vacuum of space itself as it inflates. After inflation, the universe continues to expand but at the more leisurely rate predicted by the Big Bang theory.

10^{-12} seconds to 10^{-6} seconds

After inflation, when the universe is a trillionth of a second old the temperature of the universe has fallen to 10 trillion degrees. The symmetry of the electroweak force breaks, allowing the electromagnetic and weak interactions to be felt for the first time. The universe is filled with all the particles of the particle zoo existing as virtual particles, in a state called a quark–gluon plasma. Those fundamental particles that have mass acquire it via the Higgs field. Virtual particle and antiparticle pairs come into and out of existence, turning back into pure energy when they annihilate each other. Overall there are no 'real' particles at this stage. The universe is energy, not matter, and is dominated by the strong nuclear force as suggested by the presence of gluons in the plasma that fills space.

10^{-6} seconds to 1 second

The universe is about a kilometre across. As the universe continues to cool and expand, another asymmetry reveals itself. For every 10 billion quark and antiquark annihilations a single quark remains. The Russian nuclear physicist and dissident Andrei Sakharov (1921–1989) explained it as a particular asymmetry in a single particle in the particle zoo called the K° meson. There are about 140 different mesons in the particle zoo. This tiny bias towards the existence of matter is, theoretically, enough to account for all the matter in the universe. But there is no general agreement that this theory holds up. Some scientists argue that there should be as much antimatter in the universe as matter, and the clear absence of antimatter is, for them, one of the universe's greatest mysteries. If there were galaxies made entirely of antimatter we should, from time to time, see them collide with galaxies made of matter (at which point huge amounts of energy would be released). But there is no evidence of structures in the universe made of antimatter.

The plasma of quarks and gluons begins to condense into protons and neutrons. The temperature has fallen sufficiently that quarks become confined inside protons and neutrons for the first time. The strong nuclear force that holds the quarks together, as described by QCD, has the strange property that the more a quark bond is pulled apart, the stronger it becomes, which is why quarks have never been observed in isolation. Once they became confined they are confined forever.

This period is known as the hadron era. Hadron is the collective name for all the different types of particles in the particle zoo made out of quarks. The most familiar hadrons are protons and neutrons, types of hadron called baryons, the collective name for all particles made out of three quarks. All other types of hadron are called mesons, made out of two quarks. By the time the universe is a second old, the world is full of exotic particles associated with the strong nuclear force. Particle accelerators can recreate some of these early conditions of the universe.

Also at this time, neutrinos, particles associated with the production of electrons, come into existence. They should be the most prolific particle in the universe, but so far not a single one from the Big Bang has been detected. The Italian physicist Enrico Fermi (1901–1954) first posited the neutrino as 'a desperate remedy' (his own words) required to explain the weak force. The existence of the particle was, however, confirmed in 1956.

It could be argued that the universe was already very old by the time it was a second old, even though it might seem to us, from our biased perspective, to have been inconceivably young. By that time 10^{43} Planck units of time had passed. The universe is currently 10^{60} units of Planck time old and the oldest life forms we know of appeared (a few billion years ago) when the universe was 10^{59} Planck units old. If the universe seems old in years (13.7 billion of them) then it is very, very old in Planck units, and the first appearance of life, relatively speaking, happened a moment ago.

1 second to 3 minutes

In this era, the universe is dominated by electrons and other particles that belong to the same family, called leptons. There are only six different types of lepton. The electron, muon and tau are negatively charged particles with spin $\frac{1}{2}$ but with quite different masses from each other.[5] The other three leptons are associated neutrinos, each of which is almost, but not quite, massless.

3 minutes to 20 minutes

By the time the universe is 3 minutes old, the universe begins to be dominated by electrons. It is also cool enough for protons and neutrons, via the strong nuclear force, to begin to come together in a process called nuclear fusion. The first nuclei appear in the universe. This period of nucleosynthesis only lasts for 17 minutes. After this time the temperature of the universe is too low for the process to continue. Most of the nuclei are single protons, the same things as the nuclei of hydrogen, though there are no hydrogen atoms yet. The other major constituent of the universe is made out of two protons and two neutrons, held together as a nucleus called an alpha particle, the same thing as the nucleus of helium, though again, this is a time before there are any helium atoms in the world. There are about three times as many hydrogen nuclei to helium nuclei in the universe and tiny traces of a couple of other light nuclei: small quantities of the nuclei of deuterium, an isotope[6] of hydrogen made out of a single proton and neutron unstably bound together, and tiny amounts of lithium (three protons and three neutrons bound together). And that's it: all the material that there is in the universe at this time. When the universe was a second old there were only protons and neutrons in the universe. After a few minutes the protons and neutrons begin

---

[5] Remember, there is no intuitive way of understanding what half a spin might look like. Spin is one of those peculiar quantised properties that originally bears some sort of relationship to the corresponding property in the classical world, but becomes more detached from any classical meaning as it becomes more embedded in the peculiar world of quantum physics.

[6] Isotopes are unstable atoms that have the same number of protons in the nucleus as the stable form of the atom. It is the proton count that determines the major properties of an element.

to evolve into matter that is a little more complex. Particle physics predicts that there must be seven protons for every neutron in this early universe. And this is how the contents of the universe prove to be. The predicted proportion of neutrons and protons in the early universe is spectacularly confirmed by the measurement of the quantities of hydrogen and helium existing in interstellar space today. This experimental confirmation is proof that particle physics and astrophysics describe the same reality. Here is further evidence that our separate descriptions of small and large things can be reconciled.

3 minutes to 380,000 years

Another change in the universe begins to be felt about 3 minutes after the Big Bang. The repeated annihilation of electrons and positrons (antielectrons) and other lepton and antilepton pairs creates a universe that fills up with photons (particles of electromagnetic radiation) and W and Z particles (those particles that are indistinguishable from photons except that they have mass).

After about 70,000 years the universe moves from a universe dominated by radiation to one where the relative densities of radiation and of matter in the universe are about equal.

Sometime between 240,000 and 310,000 years after the Big Bang the universe is cool enough for the nuclei of hydrogen and helium to begin capturing electrons, a process called recombination. The first neutral atoms of hydrogen and helium have appeared in the universe. Nearly all the matter in the universe is in the form of these two elements, with traces here and there of deuterium and lithium.

Before there were neutral atoms, photons (particles of light) were being continually scattered by the plasma of charged material that filled the early universe. Now that the universe is filled with neutral matter, photons are able to cohere into streams of light. The period called the Dark Ages comes to an end. A universe that was opaque becomes a universe that is transparent. The light beams flood outwards. We see this relic of the universe as it was 380,000 years after the Big Bang today as Cosmic Microwave Background (CMB) radiation: photons that have cooled from 2,700°C, when the universe was a thousand times smaller than it is now (there is a direct correspondence between the size of the universe and its background temperature), to 2.7°C above the lowest possible temperature in the universe (−277.15°C, or 0 kelvin, also known as absolute zero). This fossil evidence of the early universe is electromagnetic radiation that is so red-shifted that we see it today as microwaves with a wavelength of about 1.9 millimetres. As CMB becomes ever more accurately mapped, what we know about the early conditions of the universe becomes less speculative.

Hail the Birth of Stars

Where wast thou when I laid the foundations of the earth?

Job 38:4

A few hundred thousand years after the Big Bang the universe is more recognisably as we know it today: there is matter and there is light. An expanding universe of matter and light evolves over a period of 13.7 billion years into the universe as it manifests itself around us today.

For the universe, the past does not go away. We see what the universe was once like by looking at the light that comes to us from the universe's past – from out there. The light from distant stars convinces us that there are such objects as stars, and repeated study of the light received from many stars convinces us that the physical universe has become a hierarchy of them.

Looking out into space is the same as looking back in time. Light from the deepest past of the universe arrives as microwave radiation, the CMB, a faint portrait of the universe as it was 400,000 years after the Big Bang. It is also, in part, the faintest portrait of what we once were. CMB is a map of everything, and out of everything evolved the everything of the universe of the twenty-first century. To ask the question again – What is the universe contained in? – must provoke curious answers. We can never reach the place which is what we were 13.7 billion years ago. The universe is expanding and taking its origins ever further away. Even if we could travel at the speed of light, the horizon has a head start of 13.7 billion years. In any case, if we were to travel at the speed of light we would have to be light itself: paradoxically, it would appear as if for us there was no passing of time. We cannot see the radiation that is moving away from us at the speed of light. The furthest *objects* that are still visible are quasars moving away from us at 93 per cent of the speed of light. The edge of the universe really is a horizon; but a horizon of what, it is impossible to say. If we were physically to approach the horizon in order to see beyond it, what is beyond it need look nothing like the visible universe as we currently describe it.

The map of the early universe that is CMB looks curiously uniform, as if it is exactly the same detailed pattern repeated over and over. The pattern is homogeneous to one part in a thousand. CMB appears to confirm Einstein's assumption that the universe is smoothly distributed, at least at the largest scales it is. What the universe became is in the detail. There is enough variation there, as small as it appears to be, to account for the subsequent large structures we find in the universe as we see it today. That variation is inherited via inflation from the graininess of the quantum world. After a few hundred million years, the small amount of inhomogeneity we see in the CMB has evolved into a universe filled with clouds of hydrogen and helium molecules in a hierarchy of all sizes, which will evolve after billions of years into the complex arrangements of stars we see in the universe today.

These clouds are places of low temperature where hydrogen and helium can exist as gases. Compared with other regions of outer space they are areas of relatively high density, though they are also less dense than what we would call a vacuum in a laboratory on earth. Three-quarters of the gas in the early universe is hydrogen, and a quarter helium. Out of these clouds, stars to populate a universe will be made, which then move and evolve into familiar gravitational arrangements as clusters of stars, galaxies, clusters of galaxies, and clusters of clusters (superclusters) of galaxies.

A star is a product of turbulent gas, and gas laws (worked out on earth over several centuries) are well understood. Although the gas laws may be straightforwardly stated, how they are to be applied to these vast molecular clouds is not yet fully understood. We do know, however, that all the stars we can see being born today are emerging out of molecular clouds, and we assume that they did so in the past too.

Where there is mass, there also is gravity. Though gravity is extraordinarily weak, it now dominates the story of the universe. The science of cosmology started out as the study of the large dimensions of the universe, and was from the first largely the story of gravity. It is as if the weird weakness of gravity corresponds to the largeness of the universe that we see, as if largeness were the quid pro quo for weakness; but why this should be so is a continuing mystery.

Zooming down through turbulent clouds of different orders of magnitude, we find relatively small clouds that will condense under gravity to make a quantity of stars. Typically, for every 100 stars formed roughly 40 will be as triplets and 60 as binary systems. It is conjectured that many triplet systems eject a star and that some binary systems are pulled apart, so the solitariness of our sun is not necessarily so unusual a feature.

Gravity causes these clouds to condense and rotate. It also flattens the clouds into discs. How large individual stars eventually become depends on how dense the surrounding molecular cloud is. All stars start out with a core about a tenth of the mass of the sun that then accrete more mass from the surrounding cloud. As a star accretes mass it tends to grow more in density than in size.

The atoms of gas at the core of the star become more and more energetic as they are forced closer and closer together, which is another way of saying that the core of the cloud becomes ever hotter. The energy of the collisions causes the atoms to lose their carefully recaptured electrons to become nuclei yet again. These nuclei are 10,000 times smaller than atoms, which means that gravity can now draw them even closer together, pushing the temperature even higher in the process.

When the core temperature reaches 10 million degrees the nuclei are close enough to be fused together by nuclear fusion. The conditions are somewhat similar to how they were in the early universe when the first nuclei were being fused, except that now the process of nuclear fusion is happening in many separate pockets, each of which is the core of a star. In casual language cosmologists say that the clouds of gas collapse and ignite. Poets might hail the birth of stars. In the heart of young stars, four hydrogen nuclei (four protons) are fused together to make a single helium nucleus[1] (two protons and two neutrons) plus energy. More precisely, two of these four protons are turned into two neutrons, and the difference in energy is released as two positrons (antielectrons) and two neutrinos (particles associated with electron or antielectron production).

Mankind hasn't yet worked out how to replicate the power of the sun in a controlled way, though progress has been made over the last 50 years. Efficient man-made nuclear fusion lies some years still into the future, perhaps even another 50 years. If it is a technique we ever master, the secret of making clean(er) energy will have been found out. The waste products of fusion are helium (completely harmless) and small amounts of a radioactive isotope of hydrogen called tritium (one proton and one neutron) which has a half-life of only 12 years. Currently most nuclear power on earth comes from nuclear fission, as energy released from the process of blowing the nuclei of heavy atoms apart.

Some of the energy from the nuclear reaction taking place

---

[1] A helium nucleus is identical to an alpha particle. A stream of high-energy helium nuclei is what we call alpha radiation. A stream of high-energy electrons is what is also known as beta radiation.

in stars radiates in various parts of the electromagnetic spectrum. For distant observers it might be seen in the visible part
of the spectrum as dots of light. At this point in the life of
the universe there are, of course, no such distant observers.
How old the universe will need to be to forge such an observer
and how many observational posts there are scattered across
the universe is the subject of much speculation.

Some of the energy of the reactions is released as heat,
causing the core of the star to become even hotter. When the
temperature of the core reaches 25 million degrees the star
settles into a stable period: the gravitational pull of the mass
of the star is balanced by the force of the nuclear reaction
that is trying to blow the star apart. How long the star stays
like this, as a simple laboratory crucible manufacturing helium,
depends on the original mass of the cloud, now star.

Bodies smaller than a third the mass of the sun are too
small to shine as stars, and stars about 150 times the mass of
the sun are at the limit of how large a star can be. These most
massive stars are very rare in the universe as we see it today.
It is thought that stars over about eight times the mass of the
sun do not form out of the condensation of clouds, but how
they are formed is not yet understood. It is possible, however,
that the heaviest stars are formed in the same way as the smaller
stars.

Stars with the same mass as our sun can stay in the stable
hydrogen-burning state for billions of years. The sun has been
burning hydrogen for 5 billion years and has enough fuel for
another 5 billion years or more. Larger stars use up their fuel
faster. So stars three times the mass of the sun might be in
this state for only 300 million years. Stars of, say, 30 times the
mass of the sun will burn through their hydrogen in about 60
million years. A star a third of the size of the sun could be
burning hydrogen for 800 billion years, should the universe
last so long.

At some point, all the hydrogen gets burned up. The moment
it does, the star collapses. After millions, or even billions, of
years of stability, the star instantly collapses in on itself under

the force of gravity. The star finds a place of new stability within a fraction of a second.

As the core collapses the outer layers of the star are flung further out, making the star look huge. The star may appear to be a hundred times the size it was and red in colour. The star has become a red giant.

The core of a red giant finds a new stability at a temperature of 100 million degrees, hot enough to burn the helium that was made in the hydrogen-burning phase. A new reaction takes place in stars of sufficient size: three helium nuclei are fused together to make the first nuclei of carbon, and with an additional helium nucleus the first oxygen nuclei are manufactured.

Stars that began life 30 times the mass of the sun, and burned hydrogen for 60 million years, will now burn helium as a red giant for a further 10 million years. Our own sun is large enough to enter the helium-burning stage, where it will stay for 300 million years. All stars that start out heavier than about half the mass of the sun will become red giants. The vast majority of stars are red dwarfs, stars smaller than half the size of the sun, too dim and cool ever to achieve the yellow of our sun. There are stars smaller and dimmer still that are called brown dwarfs. These small stars do not become red giants, but merely fade and cool into black dwarfs in a process that will take many times the current age of the universe, which is why no black dwarfs will be seen for a considerable period of time to come.

Again depending on the mass of the original star, this process of collapse and increasing core temperature creates onion layers of burning in which ever heavier elements are forged. At higher temperatures carbon is burned to make the first neon and magnesium and more oxygen. At higher temperatures still neon burns, and the process continues, in burning sequences that successively use oxygen, silicon and sulphur as fuel.

Stars of about the same and up to twice the mass as the sun will only manufacture carbon and oxygen in their lifetime.

Larger stars up to four times the mass of the sun manufacture a longer sequence of elements that includes neon, magnesium and nitrogen. Stars between eight and eleven times the mass of the sun end up at the silicon-burning stage, a process that lasts a single day and out of which nickel and cobalt are made.

The cycle of burning and gravitational collapse cannot go on indefinitely. Eventually the matter in the star is as compressed as it can possibly be, according to a rule called the Pauli exclusion principle[2] that sets a quantum limit on closeness. Stars more than 15 times the mass of the sun reach this final limit. For such stars, the final burning processes run out of control and iron is made, the heaviest metal a star can forge. The nuclei of the iron atom are more tightly packed than any of the heavier elements.

The creation of these first elements locked in the heart of stars accounts for 99.9 per cent of all the elements in the universe. The universe that was entirely made of hydrogen (76%) and helium (24%) eventually becomes a universe of hydrogen (74%), helium (24%), oxygen (1.07%), carbon (0.46%), neon (0.13%), iron (0.109%), nitrogen (0.10%), silicon (0.065%), magnesium (0.058%) and sulphur (0.044%). Only 2 per cent of the hydrogen created at the Big Bang has been burned so far in the life of the universe, and the creation and burning of helium has left the amount of helium unchanged. This small proportion of hydrogen has been transmuted into eight new elements, or metals. (For some reason, cosmologists call all the products made in stars 'metals', whether any of them is actually a metal or not.) There are, however, at least another 84 naturally occurring elements in the universe, plus more than 20 elements only ever seen in laboratories on earth (and perhaps in alien cultures). These other naturally occurring elements are produced when stars of sufficient size explode.

---

[2] Named for the Austrian theoretical physicist Wolfgang Pauli (1900–1958), who made the discovery.

Stars that are several times the size of the sun have the potential to end their lives this way. After the final out-of-control burning stages, such stars have nowhere to go except outwards, and explode as supernovae in a process not fully understood. These explosions are the most massive in the universe, the energy of which brings the core elements into a series of reactions that manufactures, for the first time and in tiny amounts, the other elements seen in nature. For a while, the explosion shines brighter than a hundred galaxies. In 1987 such an exploded star was seen in the southern sky. Although it is 180,000 light-years away, for four months it appeared as bright as any nearby star. Once they have cooled, the nuclei of all the newly created elements capture electrons and become stable atoms, most of them new to the universe.

It is not entirely settled how big a star needs to be to become a supernova. Stars ten times the mass of the sun will almost certainly always end their lives in an explosion. A star a few times the mass of the sun will not necessarily explode. Those smaller than three times the mass of the sun definitely will not. For a star like our sun the far-flung layers of the red giant will eventually disperse to reveal a small dense core made of carbon and oxygen. This core is called a white dwarf and has a mass between half and three-quarters the original mass of the star. (Our sun will end up about the size of the earth but with half the sun's original mass.) This is the fate of 97 per cent of the stars in our galaxy. If the universe lasts long enough a white dwarf will gradually fade into a black dwarf. Most stars are part of a binary or larger system. The most common super-novae are called Type Ia supernovae. They are formed when a white dwarf, less than critically 1.4 times the mass of our sun, gradually accretes material from its sister star. When the mass of the white dwarf reaches this critical limit, called the Chandrasekhar limit, the star will then explode. Because this common type of supernova explodes at exactly the same limit, they all explode with exactly the same brightness. This absolute brightness is used as a candle to measure far distances in the universe. The further away such supernovae are, the less bright will they *appear* to be. Since Type Ia supernovae are so common, they can be used to measure the distance to many distant objects in the universe (because there is always likely to be such a supernova nearby).

Large stars that reach the supernovae stage (called Type Ib, Ic and II supernovae) leave behind a small dense core called a neutron star – or, for the largest stars, a black hole.

It is these exploding stars that seed the universe with the elements needed for life. It is thought that in the first billion years of star making about 500 million supernovae were created in our galaxy alone. Large stars have lived out their lives, exploded, reformed as stars and exploded again – perhaps several times over – in this time, before a small star has even had time to finish burning through its hydrogen fuel. Stars between 10 and 70 times the mass of the sun are called supergiants. There are rare stars called hypergiants, some of which are between 100 and 150 times the mass of the sun. (Whether a star is classified as a supergiant or a hypergiant isn't entirely down to mass.) The Pistol Star is a hypergiant near the centre of the Milky Way. It has a mass 150 times that of the sun, but it is 1.7 million times brighter, and has a life expectancy of only 3 million years.

*

There is striking experimental proof that this is how the elements we find on earth and elsewhere in the universe were originally forged. For technical reasons to do with the energetic properties of elements heavier than carbon, it is easy to describe how all the elements up to iron can be made in the heart of stars. The scenario that builds up the elements in the crucible of a star by fusing ever more helium nuclei together works well so long as there is carbon. What isn't easy to describe is how carbon was created in the first place. In theory it ought to be possible to forge the element beryllium from the fusion of two helium nuclei, but beryllium is highly unstable and immediately returns to the state of being two helium nuclei before another helium nuclei can be added to make carbon. In an experiment famous in the history of science, Fred Hoyle predicted that carbon could be made from three helium nuclei directly, without the intervening beryllium stage, if carbon possesses a hitherto unsuspected property. He predicted that carbon resonates at a particular energetic frequency that would allow three helium nuclei to merge smoothly together in the heart of stars, and he devised an experiment that could be carried out on earth to test his prediction. Exactly the property he predicted was discovered. Remarkably, Hoyle had found a way of testing the theory of star formation in the laboratory.

Scientists love vulnerability in a theory (particularly in someone else's theory): it provides a great opportunity to test the theory. Apparent weakness can turn out to be a way forward. It is this provisional quality of science that is often misunderstood. Science's provisional nature is its strength, not a weakness. Calling something a theory doesn't mean to say that it is merely an idea; a theory is the highest form of scientific explanation. Provisionally is how science proceeds; that is its nature.

After hundreds of years of theorising and experimentation, it is tempting to think that it should be possible by now to draw a line that encompasses some core of truth among all this provisionality. After all, every new theory must encompass what has gone before *and* describe something new. Anyone is free to say where the line could be drawn, but whoever draws such a line is not doing science, nor does the scientific method require it. Science goes out in search of greater truth (if it has to use the word truth at all) rather than *the* truth. Unfortunately, few humans can rest in such uncertainty, scientists and non-scientists alike.

Though we know a great deal about how stars must have been formed, there are many gaping holes in our understanding. The theory assumes that the first stars were made entirely of hydrogen and helium, as these were the only materials in existence, but no such stars (called Population III stars) have been observed in the universe as it is today. Population III stars may have been generally larger than subsequent stars and lived brief lives, perhaps less than a million years, before exploding as supernovae.

The oldest stars that we have found are called Population II stars. They are formed in the way just outlined except that the molecular clouds from which they condense contain, in addition to hydrogen and helium, small quantities of heavier elements forged in the heart of Population III stars and scattered across space when those first stars exploded. The addition of these extra elements may speed up some of the processes somewhat; otherwise the account of how heavier elements are forged in the heart of subsequent generations of stars remains unchanged. Our sun is an example of the youngest class of star. Population I stars contain greater quantities of heavier elements than Population II stars and are formed out of material from several previous rounds of star making.

Most of the galaxies we see in the universe today – our own is no exception – were formed early on. The oldest galaxies so far observed are about 13.2 billion light-years away;[3] that is, they were being formed about 500,000 years after the Big Bang. Until recently it was thought that galaxies were fashioned in a short period of time from vast rotating galaxy-sized molecular clouds. In this top-down description, it was argued that galaxies emerged pretty much fully formed. But recent observations suggest that galaxies have evolved out of some basic features. Early primitive galaxies are called protogalaxies, and we don't really know much about what they might have looked like. After about a billion years some of the recognisable features of galaxies as we see them today would have

---

[3] Strictly speaking they are much further away. The universe is 13.7 billion years old, and in 13.7 billion years light covers 13.7 billion light-years. But the universe is actually bigger than this because we must also take account of the fact that space itself is expanding, stretching the universe from a bowl with a radius of 13.7 billion light-years to one with a radius of around 40 billion light-years. In practice, when astronomers talk of the distance away of ancient astronomical bodies this expansion is understood, and ignored.

evolved. Globular clusters of old stars may have formed, along with a bulge of other Population II stars at the centre of the galaxy. The spiral shape of galaxies like ours may have taken 2 billion years to evolve. After this time galaxies remain relatively unchanged, as our galaxy has for billions of years now. Some galaxies are elliptical, but it is possible that they started out as spiral galaxies like ours and became elliptical as the result of numbers of collisions with other galaxies, which would also explain why elliptical galaxies are also the most massive galaxies in the universe. The early universe was smaller than it is now, and crowded with galaxies. Galaxies would have been forever bumping into each other in a way that is hard to model today.

All this makes it very hard to date a galaxy. In some sense they are all old. The oldest stars in our galaxy are around 13.2 billion years old (perhaps older), as old as the oldest observed galaxies. But it is thought that the spiral feature of our galaxy was not apparent until some time between 6.5 and 10.1 billion years ago. The age of the oldest galaxies is being continually revised upwards.

Galaxies are rarely found on their own. They are invariably part of some complex dynamic hierarchy of galaxies reflecting the fractal nature of the universe. Today, the universe is seen as filaments of galaxies with vast voids between the strands where there are no galaxies at all.[4] Where the filaments meet, dense clusters of galaxies are to be found.

A feature that all young galaxies appear to have in common is that at their heart is a massive type of active black hole called a quasar. A quasar is a black hole that is feeding on matter that strays into its gravitational ambit. As the material is drawn into the black hole it is torn apart and some of it turned into electromagnetic energy: the black hole shines brightly as a quasar. A quasar dims into a normal black hole when there is no matter

---

[4] In 2004 a huge void in space a billion light years across was discovered. It is 40 per cent larger than any other voids so far discovered, and between 30 and 45 per cent colder than the rest of space. It is conjectured that it is evidence of a collision with another universe.

around it left to eat. Such a quiescent quasar lies at the heart of our own galaxy. Outside of its gravitational influence, the shining matter of the Milky Way finds a haven.

Quasars appeared across the early universe at the heart of probably all galaxies, though how no one knows. They are an early-evolved feature of a galaxy, like Population III stars, and are not well understood. It is possible that these first quasars were formed out of the largest clouds in the early universe that were pulled together by gravity so quickly – perhaps triggered by the shock waves from the aftermath of the Big Bang ringing through the universe[5] – that they turned straight into black holes. Vast amounts of matter were removed from the universe and locked away in the heart of galaxies. The light from these old massive bodies is seen today as electromagnetic radiation that has been red-shifted to the radio wave and visible light part of the spectrum. Quasars are thought to have started out larger than the stars that subsequently appeared and by eating surrounding matter grew even bigger over time. Some of the quasars we see today are billions of times the mass of our own sun.

It is difficult to say how close together quasars were when they were formed because the universe was much smaller then. Today we see the active ones at the furthest reaches of the universe and at the centre of the youngest galaxies. The quasars that have become inactive we see at the centre of older galaxies like our own. Quasars dramatically proclaim a universe that had changed from a universe of energy to a universe of matter. About 100,000 quasars have been identified so far. How many there are altogether is disputed: there may be millions of them, or many more than that. The most distant so far discovered is 13 billion light-years away. The brightest is named 3C 273.

---

[5] In this way there is a kind of noise associated with the Big Bang. Waves of energy ring through the early universe like sound waves travelling through a medium, though the medium in this case, of course, is not air or water (a modern reconception of the music of the spheres perhaps).

It is some 2,000 billion times brighter than the sun. Although it is 2.44 billion light-years away, it is easily seen with amateur equipment. The fact that quasars can be so luminous even at such distances is an indication of how extraordinarily bright they are. It is thought that quasars must have appeared after Population III stars since there is evidence in quasars of metals heavier than hydrogen and helium.

*

Maps of the background radiation received from the early universe are getting more and more detailed. Seen in recent maps is evidence not only of CMB but also of another background radiation associated with a spectral line of neutral hydrogen. This radiation is called 21 centimetre radiation for the obvious reason that that is its wavelength. This radiation is, as yet, very hard to interpret, but already it has much to tell us of the story of the early universe. Gaps in the 21 centimetre radiation pattern appear to be evidence of a time when hydrogen atoms changed their state: from neutral atoms to a plasma of nuclei and electrons. The universe was reionised, once again filled with charged particles, several hundred million years after the Big Bang. It is in this reionised state that we see the universe today (with patches here and there of neutral molecular clouds of the right temperature where star formation takes place). It is conjectured that shock waves from the explosions of Population III stars may have caused the reionisation. The sudden appearance of quasars may have been another cause.

In spite of the lack of observational evidence for the existence of Population III stars and the uncertainty surrounding the formation of quasars, the physical description of how Population I and II stars are formed is one of the triumphs of modern physics, bringing together in a single description theories of the very small and the very large.

Here, then, for all its flaws, is a roughly stitched together route map that takes us from a state of high-energy radiation to a material world of physical bodies called stars. The endeavour is, as ever, affirmed by our ability to manufacture increasingly sophisticated technological devices, and the flaws in the theory, as ever, point the way to better theories.

The flaws can be challenging, to say the least. The failure fully to unify light and gravity at the theoretical level is also manifest in the universe at the physical level of stars. When we check the content of the physical universe using our two ways of seeing – by light and by gravity – a strikingly different picture emerges. The amount of visible mass in a galaxy or a cluster of galaxies makes a prediction about the motion of these structures. Unfortunately, the galaxies and clusters of galaxies move gravitationally as if they contain much more mass than is visible. The outer parts of galaxies move too fast, as do the outer parts of clusters of galaxies, for the amount of visible mass that they appear to contain. Our own galaxy, a flat spiral with two major arms, spinning around a central axis at 220 kilometres per second rotates faster than this at its extremities, as if there is extra mass in the galaxy that we just cannot see. It is this extra mass that seems to preserve the galaxy's spiral form. Indeed, in order to account for the observed motion of any of the large structures of the universe, each must be surrounded by a huge halo of invisible gravitational matter. Our own galaxy seems to be surrounded by a halo of dark matter ten times the radius of the visible galaxy. The disparity in the amount of mass in the universe that we see through the mediation of light and the amount that gravitational motion predicts is striking. The universe must have much more matter in it if we are to explain how the structures in it have clumped together in the way they have. There must be over five times as much unseen matter out there as there is visible matter. For obvious reasons, this unseen matter is called dark matter: 'dark' because it cannot be described by what we currently know of the nature of light. We can throw no light on dark matter. It can neither be seen nor understood. If we ever do understand it, then our understanding of light will necessarily have changed in order to bring it into view.

Dark matter hangs like an invisible cobweb strung about and across the large-scale structures of the universe. There are, of course, theories about what dark matter might be. For a time it was thought that it might be made up of massive compact halo objects (MACHOs), the collective name for all the stuff out there that would be visible if we could shine a light on it: things like dark gas clouds, dim stars (such as the possible companion to our sun), undetected planets, small black holes, and so on. But we are now certain that there is nowhere like enough unaccounted mass in the universe in this form. Weakly interacting massive particles (WIMPs) are another possibility. They are the kind of particles that are predicted by supersymmetry. The neutralino, the supersymmetric partner of the neutrino is such a candidate. But as we know, no super-symmetric partners have ever been detected.

About 9 billion years after the Big Bang, or 5 billion years ago

In the late 1990s it was discovered, much to everyone's surprise, that about 5 billion years ago the rate of acceleration of the universe began to increase. So, not only is there missing matter in the universe, but there is missing energy – lots of missing energy. This missing energy is, naturally, called dark energy.

According to Einstein's general theory space possesses an inherent energy that fuels the Big Bang. Einstein famously added his cosmological constant because he could not, initially, accept that space was not static. Once it became clear that space was not static, the constant was removed and the universe allowed to expand. In more recent times scientists have found it necessary to reinsert the constant, in order to ensure that Einstein's equations describe a universe that is expanding *faster* than general relativity predicts. (So perhaps Einstein did not blunder when he added a cosmological constant, but merely added it for the wrong reason.) This new modification gives space much the same force that gravity has, except that it is a repulsive rather than an attractive force. The nature of this force inherent in the fabric of space is apparent to us only at the largest dimensions. One attempt to explain this repulsive force posits that gravity itself becomes repulsive at those sizes. In these far reaches of the universe any matter that has not already begun to be drawn together by gravity never will be, as if we threw a ball into the air and it begins to accelerate away from us. On the largest scales expansion wins out over gravity. Galaxies and galaxy clusters are gravitationally bound but superclusters will begin to disappear over the cosmic horizon, taking the distant galaxies with them.

The size of the cosmological constant needed to describe the accelerating expansion of the universe is very small. In fact it is very close to zero, about 1 divided by 10^{60}. It seems odd that we should be in a universe where nature has chosen a number so close to, but not, zero; and this worries many scientists.

Others have wondered if dark energy is even constant, in which case it couldn't be the modified cosmological constant. There was initially some hope that the energy of the vacuum could be used to explain this acceleration in the expansion of the universe, but unfortunately the energy of the vacuum is 10^{120} times too large and would tear all matter apart in an instant. An alternative theory hopes to explain the acceleration of space by positing the existence of yet another quantum field that permeates the universe: a fifth force, sometimes called quintessence. But such a field would require the existence of yet another unobserved particle (something like the Higgs boson, but one that does not interact with matter).

These recent observations mean that we now predict that the universe will go on expanding at an accelerating rate forever.

*

If we are on the right lines with a Big Bang theory of the universe, we have to believe that 23 per cent of the stuff in the universe is matter that cannot be seen, 73 per cent is in the form of dark energy, and only 4 per cent is normal matter. Alternatively, all this missing stuff might be evidence that the Big Bang theory is breaking down. Yet the Big Bang theory has been so successful a theory in so many ways that few scientists blanch at its failure to describe most of whatever it is that the universe comprises. In any case, we don't have an alternative to the Big Bang theory, and ultimately we want our theories to fail so that we can find better ones. Finding out what is wrong with a theory is how science advances. Either the current theory gets fixed, or it gets replaced, possibly by a theory that takes a completely different approach.

Taking in the universe in a broad sweep, and putting aside for the time being the troublesome issues of dark matter and dark energy, we can assert that some half a billion years or so after the Big Bang, there was a patch of the universe that we call the visible universe seen as simple arrangements of stars, somewhat more complex than the universe of gas from which it evolved, but far removed as yet from the complexity of the world we now look out from. For all its dynamism, the rushing about of stuff here and there, we might still be inclined to think this distant universe rather boring, if this is all there is, just a number of remote fires burning in the vastness of space. This is perhaps not so terrifying a prospect after all, and yet we might be right to be suspicious of a story so simple. Is it simple because that is how it is, or because it is the only way we know how to tell the story of distant things that only look simple because they are seen from far away? An anfractuous mountain landscape, when stepped back from, resolves from jagged complexity into sinuous simplicity. The early universe is at the furthest reaches of time and space, at the horizon of our knowledge. The story may start off simple because we are only just beginning to work out how to tell the beginning of the story, or because that is how all stories must begin.

Our modern creation story is an account of how simple symmetrical structures evolve into some more complex ones, the questions being: which are the most complex, and how many of them are there? Scientists hunt around for the most complex things in the universe and try to work out how to unify the simplicity of the early universe to its later diversity and complexity. In telling the next part of the story of emerging complexity, we are forced to look more closely at what goes on inside a typical galaxy. In our search for increasing complexity we do not know where else to look. And if we are to tell that story by looking at what happened in our own galaxy, we can at least take comfort in the knowledge that we believe our galaxy to be typical of many such galaxies spread across the universe where a similar story unfolded.

Homing In

The universe has the curious property of making living beings think that its unusual properties are unsympathetic to the existence of life when in fact they are essential for it.

John Barrow

Within the visible universe that is our particular home, we are confident that none of the larger structures singles us out for privilege. We live within a typical supercluster, and the Local Group is a typical cluster that contains a typical galaxy. Nor do we believe that the region of the Milky Way that became our solar system is different from many other regions of our galaxy, or indeed of many other spiral armed galaxies, where Population I stars are to be found.

About 5 billion years ago

About 5 billion years ago in the part of the galaxy where we find ourselves today a large star-forming cloud[1] of gas

---

[1] Despite the large number of clouds of gas that do condense across the universe, compared with the overall number of clouds it is still a relatively rare event.

condensed into many stars, one of which was our sun. The patch of cloud that formed our solar system was around 24 billion kilometres across and contained materials from the lives of at least two previous generations of stars.

The hot gas in a star nursery must first cool before it can condense to form a new star. If the gas is too hot, the molecules are moving about too rapidly for gravity to overcome their motion. Indeed, gravity alone may not have been enough to have caused our sun to condense. It is likely that shock waves from a previous generation or generations of stars, together with gravity, caused the sun to precipitate.

The repeated explosion of many generations of stars makes some clouds of gas too hot for them ever to become star nurseries. They will remain ever cloud-like, as may be true of most molecular clouds now existing. Star formation has slowed right down, not for lack of hydrogen but lack of hydrogen at the right temperature, and it has come to an end in the older elliptical galaxies. The star forming days of the universe reached a peak about 10 billion years after the Big Bang and are now in a slow decline. They may be over altogether in 100 billion years time.

Gravity causes clouds of all sizes to rotate. The molecular cloud that condenses into our sun is no exception. The rotation of the cloud causes the gas in the inner part of the disc to swirl down to make an ever growing ball at the centre, and the gas and dust at the outer edges swirls ever further away. Gravity also flattens the cloud. New stars have been observed in other parts of the galaxy surrounded by just such haloes of dust. As we have already observed, how large the stars eventually become depends on how dense and how much dust is in the surrounding molecular cloud. Nuclear fusion begins when the core reaches about a fifth of the mass of the sun.

The outer edges of the cloud are cool regions, where unstable complex molecules can survive intact. When the first stars exploded, all the naturally occurring elements appeared in the universe for the first time, but so did simple molecules like water and carbon dioxide. These simple molecules appear as fine coatings of ice on small grains of dust. Some of the dust, for example, might be highly compressed carbon existing as tiny specks of diamonds or as graphite.

Cycles of star formation and explosion are a chemistry laboratory that manufactures ever more complex molecules. Hundreds of hydrocarbons (molecules made either entirely or mostly out of hydrogen and carbon) appear for the first time in star forming nebulae; formaldehyde and hydrocyanic acid and other so-called prebiotic molecules are among them. They are called prebiotic because they seem to be essential to life, but by what mechanism is still unclear. Some complex compounds found in outer space, glycoaldehyde for example, have been made to react in laboratories to make a sugar called ribose, a key ingredient of ribonucleic acid (RNA). If an oxygen atom is removed from RNA it becomes deoxyribonucleic acid (DNA).

Although the only life we know of is the life that appeared on this planet, prebiotic molecules would appear to exist across the universe. Bizarrely, these complex molecules existed before the solar system did. Some 10 to 15 per cent of the dust and gas in the molecular cloud from which our sun condensed is made out of the material from at least two previous generations of star making. Life as we know it seems to have required about 9 billion years or so of star making to produce the right conditions. And after that period of time, in many regions just like our galaxy the universe appears to be intimately attuned to the conditions required for life.

In astronomical terms, stars condense rapidly. Once the right conditions were met, our sun would have condensed and ignited within about 100,000 years, leaving behind a disc of dust that will form the rest of the solar system. The sun contains 99.9 per cent of all the mass that is available. Outside the dust cloud that shields the ignited core, temperatures are less than 30°C, no hotter than the hottest day in a typical English summer. It is in this region that the complex molecules made from many generations of star forming are protected.

We have little reason at present to suppose that the life cycle of our sun has been much different from the life of second-generation stars of the same or similar size.[2] We can convince ourselves that if we tell the story of our sun we are telling a story that is repeated many times across the universe.

The presence of carbon from previous generations of stars will speed up the hydrogen burning process slightly; otherwise, hydrogen is forged into helium in the way it is predicted to have occurred in first-generation stars. The radiation released by this reaction is carried to the sun's surface in a process that may take 10 million years, where it is released as light and heat. The sun gets lighter in both senses: it loses mass and gets brighter. And continues to do so. It will get 10 per cent brighter every billion years or so. The sun burns 4 million tonnes of hydrogen every second but given that its mass is over 10^{27} tonnes it will take, as we have already seen, at least another 5 billion years to burn through its fuel.

Only Population I stars like our sun (formed from clouds with high metal content) have planets. Before the sun has reached its final mass the planets are being formed out of what remains. The cool remnants slowly accrete over time and under gravity to form rocks in all sizes up to the size of planets. The larger particles attract the smaller ones and get bigger, like rolling snowballs. Estimates vary but tiny proto-planets called planetesimals up to 1 kilometre across take as little as tens of thousands of years to form, and those that are 50 to 500 kilometres across perhaps a few hundred thousand years.

Only a million years or so after the sun has settled into its hydrogen burning, the solar system is already a dynamic system comprising maybe 20 objects the size of the moon or larger and a million or so objects larger than 1 kilometre across, plus many more smaller objects.

Theories of planet forming are still in their early days, and theories of how the gas planets were formed are even more

---

[2] When astronomers talk of second-generation stars they mean any star that is not first generation. Some of the material in our own sun comes from a third round of star making.

tentative. Until recently it was thought that the larger satellites begin to capture by gravity the gas that did not make it into the fashioning of the sun. One of these satellites found itself at the optimum distance from the sun – where the temperature is just right – for such a process to take place. This satellite became the large gas planet we call Jupiter, taking 5 million years to reach its final mass. Jupiter's rocky core, which is 29 times the mass of the earth,[3] captures an atmosphere 288 times the mass of the earth. We do not see the terrestrial surface of a gas planet, just the top of a vast atmosphere.

Saturn struggled with Jupiter to make the next largest gas planet, taking 2 million years longer than Jupiter to reach its final mass.

As soon as the sun reaches its final mass it emits a solar wind (high-energy protons and electrons ejected from its surface) that blows away the remaining hydrogen and helium gases out of the solar system. It is conjectured that if the solar wind had been stronger the gas planets would not have formed. This is one of those worrying details that makes Copernicans, keen to preserve a lack of centrality in the universe, anxious. There is observational evidence of young stars around which no gas planets have formed for exactly this reason. Although many solar systems may have formed across the universe, we begin to wonder whether ours has qualities that make it worryingly unusual.

Because it was not so well placed, Saturn acquired an atmosphere a quarter of the size of Jupiter's, even though their terrestrial cores are much the same size. The struggle to capture gas is that much harder for the more distant gas planets of Uranus and Neptune. These four gas giants use up all the available gas.

Uranus and Neptune are beyond the frost line of the solar system and their cores are not terrestrial so much as icy, made of volatile but frozen hydrogen compounds. Further out, Pluto

---

[3] The earth is the astronomer's local unit of measurement, handy for relating different-sized objects in the solar system. Once more, aliens will have chosen a different comparative measure.

and other trans-Neptunian objects have to make do with the ice and shards that are left over, which also go to make the icy comets around and beyond Pluto that are held in the Kuiper belt, or in the far-flung Oort cloud (if it exists).

Some doubt has been cast on this theory. Observational evidence tells us that most of the large gas planets that we have discovered in other planetary systems are much closer to their suns than Jupiter is to ours. Computer simulations suggest that all the gas planets may have formed close to each other and then, because of the complex gravitational patterns between them, moved apart from each other. In this latest theory, our own gas planets may have formed closer to the sun than they are now and later moved to their present positions. In this top-down explanation the large gas planets condensed quite quickly from pockets of gas around the young sun.

Theory moves to firmer ground when it describes the fate of other rocky material that did not make it into the cores of Jupiter and Saturn. It is pulled closer to the sun, and forms the terrestrial planets – Mercury, Venus, earth and Mars – which are composed mainly of metals and minerals called silicates. This inner part of the solar system is, at this time, too hot for volatile chemicals. Assorted bits and pieces of leftover rocky rubbish orbit in a region called the asteroid belt poised between the terrestrial planets and the gaseous ones.

All the planets orbit in the same direction – anticlockwise if one were standing at the north pole of the sun – and almost in the same plane, a large-scale feature of the solar system that is unchanged from its early life as a flattened disc of rotating dust. Newton and Laplace both realised that this could not be a coincidence, and they were right. The dust rotated in the same direction when the solar system was a cloud, and continues to do so even though the dust is now contained in the heart of planets. Only a few comets travel in a retrograde direction having been buffeted into a new orbit. Halley's comet is one such.

In this description, the solar system has been reduced to a

simple dynamic system of colliding balls of matter. Earlier, the universe was clouds of colliding gas particles brought together under gravity, and earlier still a plasma of quarks and gluons. Much of the physical description of the universe seems to be about particles of different sizes hitting each other.

In regions of the universe similar to our solar system the objects are a range of macroscopic sizes and their motion is best described by Newtonian mechanics. The smaller objects in the solar system are flung about by the gravity of the larger objects, speeding them up and making them more likely to crash and fragment. The comets and planetesimals have not yet settled into their final orbits and race about here and there, knocking into each other, pushed and shoved by various gravitational forces, most prominently by the gravitational power of Jupiter. Nor have many of the large objects in the solar system settled into stable orbits yet. The comets will eventually find homes in the Kuiper belt or the Oort cloud, but on their way they sometimes bash into the planets.

Whenever the terrestrial planets are struck, they heat up. If struck enough times, or by large enough objects, they become so hot that the iron is melted out of the rocky material from which they are made. The iron then sinks to create the iron core that is at their heart. In the first 100 million years of the solar system there are many collisions and at least two very large collisions, one involving Mercury and the other the earth.

*

The earth and the solar system have been dated using various radioactive isotopes found in meteorites and from rocks collected from the moon. They are in agreement with the first accurate dating of the earth made in 1953 by the American geochemist Clair Patterson (1922–1995) of 4,567 million years with a small error bar. He used the half-life of uranium-238, which is found in some rocks on earth, to make his dating. Uranium-238 has a half-life of 4,510 million years, making it peculiarly well suited to the task. In 4,510 million years half of any amount of uranium-238 will have decayed naturally, by emitting alpha particles, into another element, in this case thorium-234. The decay is called alpha decay, mediated by the strong nuclear force. In turn, thorium-234 decays into protactinium-234 by the emission of electrons in a natural process called beta decay (mediated by the weak nuclear force). At the end of this chain of decay products, a stable (meaning a non-radioactive) element is eventually reached. The relative amounts of decay products make it possible to date the substance in which they are found.

Here, then, is another marriage between our knowledge of the quantum world (in particular of nuclear forces) and the macroscopic world (the dating of the whole solar system). The presence of uranium in the earth's core prevented the core from solidifying as quickly as it might otherwise have done. Before this effect was known of, the Belfast-born physicist William Thomson (1824–1907), later Lord Kelvin, had used the cooling of the earth's molten core as a way of calculating the age of the earth, which he incorrectly put at 400 million years, an estimate he later dramatically revised downwards. Again, only with an understanding of the properties of radioactive matter could the earth be dated accurately.

About 10 million years after reaching final mass, the earth is struck by an object the size of Mars. The impact is so violent and so much heat is produced that the iron cores of the two planets coalesce and much of the rocky crust of the earth is jettisoned into space, where it forms a ring around the earth. Over time, the ejected material is brought together by gravity and settles into a single mass we call the moon. Not everyone is convinced that this is how the moon came to be formed, but it is the most widely agreed upon current theory. Because the moon's orbit is so nearly circular it seems to suggest that the initial impact must have been almost impossibly glancing. Critics argue that a light tangential collision would be highly unlikely given the kind of dynamical state the solar system was in at that time, but a more direct impact would have pushed the orbit into a more elliptical shape than is observed. What is convincing is the fact that the moon, unlike the other terrestrial bodies, has no iron core.

The solar system begins to quieten down over time. After half a billion years the bombardment is about 99 per cent over. But then something curious happens between 4.1 and 3.8 billion years ago. The bombardment picks up again during a period known as the late heavy bombardment. The number of craters on the moon that date from this time is evidence that the solar system had once again become a violent place. It isn't known why this late bombardment occurred, though it is conjured that if the major gas planets did indeed move into their current places from elsewhere then complex gravitational tides would have upset what had otherwise become a stable solar system.

By the end of this period, the universe had returned to some sort of equilibrium and became much as we see it today. There are just a handful of major orbiting bodies; everything else is now stored away in the asteroid belt, Kuiper belt, or Oort cloud. By this time the solar system has become so stable – excepting the occasional catastrophic event – that it seems unlikely that it will change much in the next few hundred million years, unless mankind finds a way to upset the balance. We live in an environment where the probability of being struck by a major comet has been reduced to once every 10 billion years, and by a small one to once every 10 million years. Within these long odds of astronomical upset we frame an illusion of stability.

*

We are beginning to find evidence to support the idea that our solar system, too, may be typical of other planetary systems across the universe. Until 1992 the solar system was the only multiplanetary system we knew of. In that year, a second planetary system was discovered, around a pulsar labelled PSR 1257+12, some 980 light-years from the sun. In 1995 a Jupiter-sized planet was found orbiting a sun-like star named 51 Pegasi. In 1999, a multiplanetary solar system was discovered in which the sun is in its hydrogen-burning sequence, as our sun is. The planetary system is in orbit around the primary star in a multiple star system called Upsilon Andromedae, about 44 light-years away. And so astronomers begin to grow more confident that they will find many other solar systems, and solar systems that look increasingly like ours.

Several other jupiters, by which is meant a large gaseous planet like our own Jupiter, have been discovered, now that we have the technology sophisticated enough to allow us to see what was always there. Our Jupiter is over twice the mass of all the other planets added together. The current technology has now crossed this jupiter limit, and new techniques and improved technology will doubtless bring us ever closer to our future hope of finding other earths. In particular, we are looking for what are called 'Goldilocks earths',[4] where the conditions are just right for life.

The first terrestrial planet seen outside our solar system was discovered in 2005, and since that date other terrestrial planets have been identified. Called superearths, these terrestrial planets are at least five times the mass of the earth, more like our smaller gas planets but without the gas. In 2007 the first Goldilocks earth may have been detected. There are three planets orbiting the star Gliese 581, about a third of the mass of the sun, and 20.5 light-years from here. The planets appear to be earth-sized.

The first requirement of finding another example of something, whatever it might be, is that we have understood what that something actually is. It doesn't take much to recognise another ball when we see one, but to understand what another earth might look like, we must first know what appears to make this earth look special; only then will we know what to look for in another earth. It is the desire to undermine this appearance of privilege that has driven science forward

---

[4] Goldilocks was fussy. The temperature of her porridge had to be just right: not too hot and not too cold.

in the 400 years since Copernicus removed the earth from its position at the physical centre of the universe, and unwittingly established the scientific principle that not only is the earth *not* at the centre of the universe, it is not central to the universe in any way at all. The discovery of similar conditions on other earths will encourage us in our Copernican belief that what happened here not only happened over there, but perhaps in many places in the universe. The discovery of other earths makes it possible to turn our earth into an experimental object that can be compared to others of the same type. The existence of other earths will enrich our understanding of the differences that either make our earth special, or not special. Whatever specialness there is becomes ever more refined.

For now it is still possible to hold on to the belief, should that be one's inclination, that there is just the one earth. Conversely, the determined belief that the earth is not unique pushes the materialist endeavour further forward, with the implication that any windows of opportunity that proclaim the earth as unique are getting smaller.

Once more, we set off with a determination to undermine our own supposed uniqueness, not necessarily with the intent of showing up our insignificance, but because this is one way of formulating the scientific quest, as the desire to understand more deeply what it is that makes us what we think we are. For the moment, the narrow focus that directs our attention to this third rock from the sun throws up all sorts of troubling, or stimulating, examples of privilege that science must try to address. The undermining of such privilege only moves the search elsewhere. Whether or not such an enterprise will ever come to an end must be a matter of belief.

The obvious standout quality of the earth is that it is home to life. As Copernicans we are convinced that there must be life elsewhere, but before we can go in search of such life we must first figure out what we think life is, and what conditions it requires.

If Jupiter had not been there to protect the earth from bombardment, it is hard to imagine how life on earth could be possible. And here already we meet our first challenge. It is not just the conditions on earth that appear to make life possible, but the conditions of the solar system itself seem peculiarly necessary for it. We also know that the conditions of the universe as a whole may make life rare, given that it seems to require about 10 billion years of star formation to make the right sort of molecules for life, and the fact that star forming itself seems to have slowed right down. Such arguments that relate the conditions of the universe to the conditions of mankind are illustrations of the anthropic principle, used by Copernicans and non-Copernicans alike.

The anthropic principle is a useful way of trying to assess how much leeway there is in the parameters that we think must have determined how complexity arose in the universe. The anthropic principle could be used, for example, to account for the peculiar flatness of space or the near vanishing smallness of the cosmological constant. A Copernican might argue that the extraordinary flatness of space encourages us to believe in the existence of other universes in which space is curved in all sorts of different ways, and in which there are either no observers or observers quite different from us. We observe a universe of such precise flatness because only in such a universe could observers, such as we are, have evolved. This is one way – rather tortured perhaps – of protecting the idea that we are not privileged. Parallel universes and the multiverse are ways of preventing the quantum laws from appearing to be inextricably linked to our own existence as observers of the universe. A non-Copernican, on the other hand, might argue that there is very little wriggle room, and use the same anthropic principle to eschew the profligacy of other universes and to assert the scarcity of complexity. The second law of thermodynamics tells us that any system must become less ordered over time. The universe can create order only at the expense of less order elsewhere. The question is, what is the cost of our assembly as humans? Dare we believe that the cost is a universe precisely of this size and energy?

The chance strike that knocked off a part of the earth and made the moon, also seems to be necessary to the presence of life as we know it. The moon stops the earth swinging wildly on its axis, reducing this wildness to a wobble. Without a large moon the earth would topple over, even more dramatically than Mars does. The modest wobble of the earth, taken together with the inclination of the earth to the sun, accounts for the moderate change in the seasons, which would otherwise be too dramatically changeable to support the kind of complex life found here. Without the ameliorating presence of the moon, life would have to have taken a very different form, which is not to say that life is ruled out, just that we cannot yet imagine what a different sort of life might look like. How we describe the universe is ultimately limited by the power of our imaginations. Given that we are part of the output of the universe, it does not seem likely that we could ever be more imaginative than the universe we attempt to describe. Whatever we think the universe is must always be at the limit of our ability to imagine what it could be.

Sometimes scientists come across coincidences that are so bizarre it is difficult to know how to classify them. When the moon was first formed, it was a third of its current distance from the earth and so a lunar month only lasted about five days. The moon has been receding from us by 38 millimetres a year, and in consequence slowing down the earth's rotation.[5] Today the moon has receded to a distance where it so happens that it is 400 times closer to the earth than the distance from the earth to the sun. This might not seem so remarkable except for the fact that the moon happens to be 1/400 of the width of the sun, which means that during an eclipse the moon can cover the sun exactly. This will not happen in the

---

[5] Some time ago the earth, in turn, slowed down the moon's rotation so that the moon and earth are locked together, the moon always showing the same side. Actually, the moon, like the earth, wobbles, so it shows slightly more than one side. Eventually the moon will return the compliment and the earth always show the same side as seen from the moon.

far future and did not happen in the far past. The ancients used this fact to come up with the first estimates of the distance to the sun. It seems unlikely that we can make anything of such a coincidence.

Because the earth has an iron core and is spinning, it generates a magnetic field that protects it from the damaging effects of radiation – damaging, that is, to the kind of life that exists here. Cosmic rays, emitted by the sun as winds of protons and electrons blowing at 400 kilometres a second (three times faster in a solar storm), are deflected by the earth's magnetic field. Astronauts have to be very careful to avoid these dangerous rays when they leave their spacecraft. If life must have this sort of magnetic protection, then we must hope to find other earths with magnetic fields. Alternatively, we have to think up other ways that complex life could have evolved that is unharmed by high levels of radiation.

Were it not for the strength of the earth's magnetic field, cosmic rays would also have stripped away the earth's atmosphere. Mars has no atmosphere because its magnetic field is too weak. Whatever life-forms there are on earths that have a weak magnetic field might also have to manage without an atmosphere.

Even the amount of uranium in the earth seems to be perfectly balanced for life. Too little and the earth would have cooled too quickly. It would have become something inert. Too much and the radioactivity levels would again have made this sort of life impossible. The level that we do have suggests that the sun is made of material from a third round of star making, which again reminds us that not only are the conditions of the solar system finely balanced for our life, but the conditions of the universe are finely balanced too.[6]

---

[6] The Hungarian-born American mathematician John von Neumann (1903–1957) discovered an even more peculiar relationship between mankind and uranium. He remarked: 'If man and his technology had appeared on the scene several billion years earlier, the separation of uranium-235 [crucial for

Once the solar system had settled down, and the earth was protected from the devastation and reversal of frequent cataclysmic collisions, smaller collisions with ancient rocks called chondrites allow the story of increasing complexity to unfold further. The unfolding story of complexity is still about balls of matter hitting other balls of matter. Gas particles became stars; galaxies collided with each other to become the universe as we see it at its largest sizes. Within galaxies we find solar systems that behave like balls on a billiard table. Now we look more closely within one stable solar system – ours – where smaller impacts tell a new story of emerging complexity.

Chondrites bring to earth the complex molecules that once floated in the cool distant reaches of the cloud of dust from which the sun precipitated. Chemicals older then than the sun and forged in generations of stars are brought to the earth as seeds of life. Chondrites fall to earth even now. One fell on Murchison in Australia in 1969 and was found to contain 411 different organic compounds, including 74 amino acids, eight of which are found in proteins of living organisms. A study of relative abundances made by Armand Delsemme in the 1970s shows that there is a strong correlation between the abundances of hydrogen, oxygen, carbon, nitrogen and sulphur in living organisms and in material found in comets. Life betrays its cometary origins. Phosphorus is the exception, found in all living organisms (though only in one molecule) but not in comets. Conversely, among the cosmic abundances life has found no use for just one element: inert helium. Life, for all its complexity, is woven out of only 30 or so different molecules, constructed out of the most abundant elements in the universe.

making atom bombs] would have been easier. If man had appeared later – say 10 billion years later – the concentration of uranium-235 would have been so low as to make it practically unusable.' There appears to be a fine balance between when we discovered the means to annihilate our species and how smart we needed to be to make the discovery. Weighed in that balance is the unanswered question: Are we smart enough not to annihilate ourselves?

It seems likely that life on earth did not begin from scratch. Sophisticated molecules that appear to be built into life were first formed in outer space long before the earth existed. When we wonder where we are to find alien life, it may be that we are looking in the wrong place. We *are* alien life. We came from out there, and there may even be other alien life (bacterial probably) here on earth that we have not yet discovered.

Scientists talk of the 'classical habitable' zone for life, which in many respects is disappointingly (or intriguingly, depending on where you have put your faith) narrow for the earth. The earth is placed at such a distance from the sun that water can exist as a liquid. In fact the earth is the only place we know of in the universe where water exists in all three states: as ice, water and steam. The only place we know of *so far*.

Water (H_2O), the most common triatomic molecule in the universe, was also brought to earth by chondrites and in cometary dust. At least 30,000 tonnes of water arrives on earth as cometary dust every year, even now.[7] At some point in the earth's deep history the atmosphere fills with water vapour and it begins to rain for the first time: torrential rains that fill the oceans. The oldest fossil evidence of rain is found as indentations in rocks discovered in India that are at least 3 billion years old, but it is thought that it had already been raining on and off at least a billion years by then. Even the properties of water seem to be finely tuned to the possibility of life. It has been suggested that the peculiarly complex quantum bonds in water might be connected to life. More mundanely, if ice wasn't less dense than water – and it is unusual for a solid to be less dense than its liquid form – the oceans would have frozen from the bottom up and killed all marine life.

We all know that water is essential for the existence of organic life: 'Without water, it's all just chemistry,' says Felix Franks of the University of Cambridge, 'but add water and

---

[7] Not a huge amount of water: enough to fill a pool 100 metres long, 30 metres wide, and 10 metres deep.

you get biology.' It is less obvious that it is also essential to the inorganic life of the planet.

The landmasses of the earth have shifted and changed over the ages and it is the tectonic plates that carry them. If it were not for water the tectonic plates would not move. Water acts like the oil in the machine that moves the continents around the globe. Currently there are seven major plates and many minor ones. We don't know how the land was distributed among the oceans when the earth was very young, but we do know that the face of the earth has looked very different over more than 4 billion years of history. The plates are made of two upper layers, the crust and the lithosphere, which together move slowly across a lower layer called the asthenosphere. The plates can move between 0.66 and 8.5 centimetres a year, not so different from the growth rate of a toenail. The illusion that the solid plates move like a liquid is due to a process called creep, in which the mineral grains that make up the lithosphere continually re-form in one place as they are removed from the other, giving the impression of forward motion. Because of tectonic action there is no longer any evidence on land of the early bombardment of the earth, but there is evidence under the oceans, and also on the cratered surface of the moon which has no tectonic plates.

The movement of the earth's tectonic plates is currently widening the Atlantic Ocean, sending Washington DC and Paris away from each other by 30 centimetres every ten years. Correspondingly, the Pacific Ocean is shrinking. The most travelled part of the modern world is the panhandle of Alaska, which was once attached to what is now eastern Australia. It broke off 375 million years ago – quite recently given that our current focus is on the last few billions of years – to begin its journey northwards. Other landmasses might have travelled even greater distances in the deep past, but this history is, at least for now, lost to us.

Along the boundaries of the tectonic plates, violence happens: earthquakes and volcanic activity. It is here that mountains are raised and ocean trenches are dug. In a thousand years, in the world as we see it today, the Himalayas can grow in some parts by a metre and be eroded in others by more than a metre. In the deep past, of course, the Himalayas did not exist.

If the earth were suddenly to lose its water it would become more like Venus is now, a place with tectonic plates that no longer move. Venus had a violent tectonic past, and if its atmosphere ever changes may have again in the future. Currently the atmosphere of Venus is mainly carbon dioxide that has locked the planet into an extreme greenhouse climate. Carbon dioxide lets through light except for the part of the spectrum around the infrared region. The solar radiation that gets through hits the ground and is radiated back as the missing part of the spectrum, that is, in the infrared region. The infrared radiation cannot escape for the same reason that it originally could not get in: the presence of carbon dioxide in the atmosphere. Since infrared radiation is heat, it is heat that becomes trapped within the planet's atmosphere. Glass has the same property: it filters out the infrared region of sunlight, which is how greenhouses stay warm and why this process is called the greenhouse effect.

The greenhouse effect has raised the surface temperature of Venus to 400°C. At the other extreme, Mars is a cold desert planet without the molten magma that fuels volcanic action. The earth is poised between these two landscapes, and it is the atmosphere that preserves this delicate balance. Bizarrely, if the formation of the moon had not removed a large part of the earth's surface, the surface would have been too thick for tectonic movement to have occurred. Again, the moon is a significant part of the story of why there is life on earth.

The earth's gravitational field is just the right size to hold an atmosphere, as is its magnetic field. The moon has no atmosphere because gravity is too weak. As close as it is, the moon is an inhospitable place. Temperatures fall to -170°C at night and rise to 100°C during the day. The earth's first atmosphere was mostly made of the hydrogen that was released when the earth began to cool around 4.3 billion years ago.

This atmosphere changed with the addition of gases pumped from the volcanoes: ammonia, methane, carbon dioxide and water vapour.

When the earth was young the sun was a third less bright than it is today, but because of the high levels of CO_2 and the heavier atmosphere, the ground temperature was $100°C$ and the oceans were close to boiling.

Though the magnetic field protects the earth from the worst effects of radiation, we have become more protected in modern times by the complexities of an evolved atmosphere. The atmosphere is a series of shells within shells: the magnetosphere, exosphere, ionosphere, mesosphere, stratosphere (which in turn contains the ozone layer) and troposphere. The solid earth is another series of layers: crust, upper and lower mantle, and a molten iron layer that surrounds a solid iron core. The ionosphere, the upper layer of the atmosphere 80 kilometres above the surface of the earth, absorbs X-rays and some ultraviolet radiation. The ozone layer, 20 kilometres above the surface, is made of unusual oxygen atoms, O_3 rather than O_2, which are particularly adept at absorbing ultraviolet radiation. The heavier oxygen molecules are a product of the dissociating effect ultraviolet radiation has on water molecules. The ozone layer began to build up early on in the earth's existence, though there is no ordinary oxygen on the earth at this point.

The earth was fashioned out of rocky material that accreted from a disc of space dust, but there are no such rocks in existence on earth today. This primordial material has been turned by actions of the earth as a whole into the rocks with which we are familiar. The volcanoes turned the crust into various forms of rock called igneous rocks, the two main forms of which are basalt and granite. Basalt is lava that cools rapidly as it pours out of volcanoes and makes the ocean floors. Granite is lava that cools slowly deep underground and lies under most of the continents.

It is hard to tell how old the oldest rocks are. The violent collisions of the first half billion years of the earth's life repeatedly melted the surface of the planet and set the geological clock back to zero. The best we seem to be able to say is that rocks are as old as it is possible for them to be in the form in which we find them. Geological time begins with these first earth-made rocks. The oldest igneous rocks on the planet have been found in Canada and are thought to contain grains of rocks over 4 billion years old, which is why we know that water must have been around for at least as long, since tectonic action requires the presence of water. The best place to find old rocks is on the moon, where there is, of course, no tectonic action. Some moon rocks have been shown to be some 4 billion years old.

So much for some of the physical conditions of the earth, that may or may not be prerequisites for the emergence of life elsewhere. But before we can begin to decide on the necessity of those conditions for life as we know it, we must work out how the physical earth became a living earth in order to understand more fully what we even mean by life.

Life, like the *Mona Lisa*, may be 'older than the rocks among which she sits'.[8] Controversially, it has been suggested that some simple bacterial life first arrived from outer space, perhaps brought to earth by some passing comet. Microbes could have hitched a ride on ancient rocks. Less controversially, it is thought that during the first few hundred million years of the earth's history life may have started over several times, only to be killed off by the violence of the environment at that time. Life, like every other emerging form of complexity that the universe unfolds, has no choice but to emerge at the first opportunity, and to persist so long as the conditions remain favourable. Whatever it is, and however it got here, as soon as the coast is clear *life* emerges.

---

[8] Walter Pater's famous aesthetic judgment of Leonardo da Vinci's master-piece in *The Renaissance* (1893).

Beginning the Begats

Now I am ready to tell how bodies are changed
Into different bodies.

Ted Hughes, *Tales from Ovid*

Je suis un ancêtre.

Napoleon

Some grand Americans trace their ancestors back to the *Mayflower*. And there are those who say it is grander still to have ancestors on the ship that followed, the servants having been sent on ahead. As anyone who has ever tried to put together a family tree knows, it is hard to fill in an unbroken line of family descent going back even a few centuries. Posh English families search for relatives who might have come over with the Normans, but even such rare finds trace a family tree a mere thousand years old.

Six hundred or so generations would take us back to the first farming communities of around 10,000 BC. But no one has traced a family history back that far. The Bible, in one of its more tedious passages, has a go at tracing a line of descent from the ancient Hebrew tribes, and for a time this was the way that some cultures worked out how old they were. In the early seventeenth century, reports from China had begun to tell of an emperor dating from 3000 BC and of a history perhaps much more ancient than that. Similar tales were coming out of India, seeming to suggest that these civilisations were as old if not older than Hebrew civilisation. It was this disturbing prospect that led Newton to devote so much of his own time to working out the ancestry of the families of the Old Testament. A generation later, the French writer and philosopher Voltaire (1694–1778) claimed that Eastern cultures were superior, a heretical stance that the Church tried to undermine by undermining Voltaire's reputation, although, contrary to popular belief, Voltaire was not an atheist, merely opposed to organised religion.

In his *Annals of the Old Testament*, published in 1650, the Archbishop of Armagh, James Ussher (1581–1656), had worked out a chronology of Creation. In a supplement to this work published in 1654 he calculated that Creation had occurred on the evening before Sunday 23 October 4004 BC, a date that does not differ much from the attempts of others, from at least the time of the Venerable Bede (c.672–735), to set a date for Creation. Ussher is today often taken for a fool, but he was a greatly respected scholar of his time, known throughout Europe. According to some biblical scholars, the reign of man was meant to last no more than 6,000 years, taking as evidence a line from the Book of Peter: 'One day is with the Lord as a thousand years, and a thousand years as one day' (2 Peter 3:8). Creation, which began around the year 4000 BC, was set to end 6,000 years later. Today, we believe that in 4000 BC the wheel was being discovered in Mesopotamia. Ussher's date was inserted into the margins of editions of the *King James Bible* from 1701. It is to this version of the Bible that fundamentalists have their curious relationship.

By Newton's time some believers knew that the earth must be much older than this timing would suggest. Newton thought the earth might be as old as 50,000 years, and the French naturalist Georges Buffon (1707–1788) hazarded 70,000 years. In the mid-eighteenth century, Kant wondered if the earth could be as old as a million years (and the universe myriads of millions of centuries old). The French mathematician and physicist Joseph Fourier (1768–1830), through mathematical analysis of heat loss, estimated that the earth must be about 100 million years old. We now know, because of what we know about the universe at its smallest and largest dimensions, that the earth is some 4.6 billion years old, and we have only known this since the 1950s.

If the paths into historical time are ill defined, how much more indistinct are the paths into geological time. Organisms live and die and mostly disappear forever. We are lucky that we can trace any path at all through the geological past of hundreds of millions of years. Once we leave history and step into the deep time of evolution we are forced to confront the fact that apparently all that survives to guide us is a sparse collection of fossils; and it is remarkable that there is even that, the conditions required for their formation being so extraordinary. An organism with a skeleton must be caught by death in places where slow decay can occur, and where mineral deposits can be laid down in a process of sedimentation that gradually replaces the minerals of the organism's skeleton to make an almost perfect copy. Even more rarely, the soft parts of animals and plants may be captured by sedimentation, or in tree resin that has somehow managed to hold a small creature or a part of a plant before hardening and fossilising into amber. Without these rarities it is hard to imagine how we could ever have begun to prove a theory of evolution. Fossils have been found in Etruscan burial chambers, so were presumably known to be significant from the beginnings of civilisation.

Given that there are so few fossils, how can we ever hope to say which might be an ancestor? We have to give up the idea that we can ever make a direct connection between something living now and the fossil of something that once lived in the deep past. Extinction is the rule of evolution and survival the exception. Not only are fossils extremely rare, life has been unbelievably abundant. We see it in nature as it manifests itself around us, in frogspawn or seeds, but we can hardly begin to imagine the abundance of deep time. The idea that we could trace a line of direct descent through the billions of organisms that have ever lived is, if not theoretically impossible, clearly impossible in practice.

Evolution is the conviction, as expressed by the English naturalist Charles Darwin (1809–1882) in *The Descent of Man* (1872), that all life can be traced from 'some less highly organised form'. It is, says the American philosopher Daniel Dennett (b.1942), 'the single best idea that anyone has ever had'.[1]

All living forms have common ancestors, and ultimately a single common ancestor that lived in the deepest reaches of the geological past. Instead of looking for a direct path of *descent*, and despite the title of Darwin's famous book, the theory of evolution fumbles in the dark of deep time in search of common *ancestors*. If we believe that there is a single common ancestor from which all life evolved then we must also believe that it is possible to arrange all descendants into a hierarchy of relatedness that converges on this single ancestor running backwards through geological time. We'll never find the single common ancestor – it was all too long ago – but it turns out that it is possible to work out how closely related we are to the living, and to anything that once lived.

The naming and relating of different living forms did not begin with the theory of evolution. In the eighteenth century living things were being classified by the Swedish naturalist Carl Linnaeus (1707–1778) into genus and species. Even then the idea of classification was not new,[2] but no one had ever been as systematic as Linnaeus. He organised and named around 7,700 species of plants (by concentrating on differences in their sexual organs) and 4,400 species of animals. For the first time, humans were included in a classification alongside other life forms. But Linnaeus held to the orthodoxy of his time, that species are 'as many . . . as originally created by the Infinite being', that all species had existed unchanged from the Creation, when they had come into existence in one fell

---

[1] In Dennett's book *Darwin's Dangerous Idea* (1995).
[2] In the sixth century BC Anaximander guessed that life had begun in the sea because of the visible structural resemblances between man and fish.

swoop. Linnaeus thought that apes were humans with tails long before the time of Darwin but there was no concept of change, of evolution, in Linnaeus's worldview, rather the opposite. Before Darwin, life was seen as a fixed 'great chain of being' ordered from the lowest to the highest forms with man at the pinnacle. In 1837, a mere 22 years before *On the Origin of Species* was published, the British philosopher William Whewell (1794–1866) insisted that 'species have a real existence in nature and transition from one to another does not exist'.

It was this notion of the fixity of species that the Darwinian revolution overthrew and which put science once again in direct conflict with the Church. Man is neither the purpose of Creation nor its end point. Purposefulness becomes redundant in the face of the random winnowing of nature and time. Animal life is not arranged in a hierarchy of inferiority to man for his exploitation: man is an animal too. Darwin's principle removes mankind, yet again, from a position of privilege, which makes evolution another form of the Copernican principle. Man emerges bloodily and by chance from among brutish animals, indeed ultimately from slime.

The idea of a single Creation had been coming under fire for some time before Darwin. In order to account for the fossil evidence of once-existent life forms and the lack of such evidence of current life forms, the French naturalist Baron Cuvier (1769–1832), the leading naturalist of his day, had put forward the idea that there had been many (32 in fact) extinctions and creations (though this was an idea he was later to abandon in favour of the orthodox view that species are unchanging). Others, in an attempt to save the biblical Creation story, took fossils to be evidence of animals that had perished before the Flood, a theory that, before there was a theory of tectonic uplift, seemed to be supported by the fact that fossils were often found on higher ground.

Baron Cuvier also advanced the study of comparative anatomy, comparing living and fossilised species. It was said that he could reconstruct an animal from a single bone. Organising the animal kingdom by common features was well under way before the time of Darwin. We are now convinced, for example, that it is a peculiar feature of the middle ear, among other particularities, that collects dogs, foxes, bears and racoons together into a single family, Canidae. A particular sort of tooth called the carnassial is one of the features that characterises a diverse order of placental mammals called Carnivora. Confusingly, not all species belonging to the order Carnivora are carnivorous. Pandas, for example, are almost exclusively herbivorous. And conversely, not all carnivorous species belong to Carnivora. Humans are often carnivorous but do not belong to Carnivora. But all Carnivora are mammals, as humans are. Birds, lizards, snakes and turtles are not mammals but they share with mammals their development out of an egg that has protective amniotic fluid. Vertebrates is a larger classification that includes amniotes and all other animals with a backbone. Arranging organisms by visible and shared structural features is highly suggestive that there are shared common ancestors, but such an arrangement does not prove evolution no matter how suggestive, nor does it provide an explanation of how evolution happens. The study of embryonic development also hints at descent from some common ancestor. The gills that appear at one stage of the embryonic development of mammals are a clue to our common ancestorship with fish. Or we can relate the way the heart works in mammals to its origin in fish, where it follows the same basic plan.

Charles Darwin was not the first to suggest that evolution happens. His grandfather Erasmus Darwin (1731–1802) had wondered if all animals were descended from 'one living filament', and Georges Buffon, more particularly, if the North American bison was descended from an ancestral form of the European ox. The French naturalist Jean-Baptiste Pierre Antoine de Monet, Chevalier de Lamarck (1744–1829) was an early propounder of the theory of evolution but his explanation has not found favour in modern times. Lamarckianism is the idea that characteristics developed in a lifetime – a muscular body resulting from frequent visits to the gym, for example – can be passed on to future generations. Darwin had a different explanation of how change passes from generation to generation. Nature selects characteristics most suited to a changing environment. In tandem with his idea of natural selection, he added sexual selection: the two principles that describe how evolution works at the macroscopic level. The drab female bird uses sex as a means of controlling natural selection. The female chooses whom among the males gets to breed and who doesn't.

Sexual selection often appears to work against natural selection. The large and brightly coloured tail feathers of the male peacock hamper flight and make them vulnerable targets, but within the environmental niche that peacocks inhabit, and that has been shaped by natural selection, this sexual extravagance comes as a sort of luxury of complexity.

Darwin reduces nature to randomness, purposelessness, sex, violence and death. In a different field, and a few decades later, Sigmund Freud (1856–1939) would come to a similar reduction.

The idea of natural selection, its random amorality – 'murder and sudden death are the order of the day'[3] – had a huge impact on Victorian society. The notion that only the fittest survived led to the unbridled capitalism of Social Darwinism, a philosophy developed by the English economist Herbert Spencer (1820–1903), who coined the phrase 'survival of the fittest' and invented laissez-faire economics. That species were perfectible through breeding spawned the science of eugenics, first propounded by Darwin's half-cousin the polymath Francis Galton (1822–1911). Darwin was troubled by the bleakness of his own vision: 'It is hard to believe in the dreadful but quiet war going on in the peaceful wood and quiet fields.' He eventually lost his faith in God, replacing it with a stoicism that found comfort in contemplation of the majesty of the universe.

Darwin believed that, gradually and smoothly over long periods of time, natural and sexual selection change one species

---

[3] The words of the English naturalist T. H. Huxley (1825–1894): known, because of his fierce defence of the theory of evolution, as Darwin's Bulldog. He was, perhaps, the Richard Dawkins of his day.

into another species. Darwin argued at length that artificial selection as conducted by mankind in, say, the breeding of dogs and cats can be taken as evidence in favour of natural selection. In just a few generations mankind can breed very different-looking creatures. Natural selection operating over vast tracts of time could effect even more powerful change. But Darwin's arguments don't amount to a proof. Artificial selection could just as well be used as an argument against natural selection. In artificial selection mankind gets to choose what lives and what does not, which is no different from the kind of intervention that some all-powerful god might make. Nor is it encouraging that prominent characteristics selected by breeders often bring with them some inherent weakness: in Pekinese dogs it is breathing problems, in blue-eyed white-haired cats deafness, and so on. A deaf cat and a wheezing small dog would be at decided disadvantages in the wild. Artificial selection shows the problems that arise if the changes made are too great.

If, as Darwin argued, natural selection makes *tiny* changes over very long periods of time then it becomes necessary to prove that the earth is old enough for evolution to have taken place. Natural selection also lacked a mechanism, without which Darwin's theory would be useless. And there was a further problem: no fossils had been found of any of the transitional forms between one species and another that would be evidence of gradual change.

The question of time was beginning to be addressed by the rising science of geology. James Hutton (1726–1797), a Scottish farmer, had established a notion of gradual change, called uniformitarianism, within geology. He'd noticed that Roman roads, although they had been built some 2,000 years ago, were still visible. Erosion had taken place, but slowly. It led him to speculate that slow processes of erosion and sedimentation of rocks could be used to measure out geological time. Even though these uniform geological processes are interrupted by violent events like earthquakes and volcanic eruptions, so long as we understand that the underlying processes are relentless and uniform, then this uniformity could be used as a sort of clock. Hutton's principle made it possible to date sedimentary rocks for the first time. And by dating the rocks, the fossils contained in them could also be dated. His ideas were popularised by the Scottish geologist Charles Lyell (1797–1875) in his three-volume work *Principles of Geology*, published between 1830 and 1833. It was in these works that Darwin was first introduced to Hutton's influential ideas.

Darwin's weak argument to explain the lack of intermediary forms in the fossil record was that the intermediary forms between species were extinct, and because the fossil record was so sparse and nature so profligate they had been lost to view. Despite the title of his other famous book, Darwin didn't actually know how to explain the origin of species. Rather, he describes the evolution of complex forms, which *appears* to happen by natural and sexual selection. But without a mechanism to explain *how* it happens Darwin's beautiful theory would ultimately have to fail. For all its initial impact, Darwinism went out of fashion.

In the late nineteenth and early twentieth centuries, the almost simultaneous rediscovery by three biologists of the work of Gregor Mendel (1822–1884) seemed to cast further doubt over Darwin's theory of evolution. Mendel became an Augustinian monk, and between 1856 and 1863 in the monastery at Brno he bred some 29,000 pea plants in a plot of land about 35 metres long and 7 metres wide. He analysed some 13,000 of them, recording what traits were passed on from generation to generation. Mendel's data seemed to suggest that there is some sort of fixed quantity (or 'quantum') of inheritance, an idea at odds with Darwin's smooth theory of gradual change. This conflict might remind us of the problems there have been trying to reconcile the jagged theory of quantum physics and the smooth theory of general relativity. In retrospect, one of the clearest indications that breeding does not blend characteristics is apparent in the simple fact that out of a male and a female there invariably arises a male or a female, and not some blended hermaphrodite. That there are discrete things in the pea plant that determine its overall properties was the first step towards the discovery of genes.

A synthesis of the two mechanisms of evolution – the inheritance of variation (rather than blending) and natural selection – was made in the 1930s and 1940s by, among others, the American biologist Sewall Wright (1889–1988), and the British biologists J. B. S. Haldane (1892–1964) and Ronald Fisher (1890–1962). Darwinism was reclaimed about 90 years after the first publication of *On the Origin of Species*. The scientists that forged the modern evolutionary synthesis showed that the two theories are not incompatible. A spectrum of possible variation is described by the new science of genetics, and natural selection ensures that those variations most fitted to the environment are preserved. The spectrum of variation is not continuous, as Darwin thought, but made up of discrete bands. At the microscopic level that discreteness is the genes, which we now know to be separate packets of information housed on a very long and complex molecule called DNA, contained in each and every cell of all living things.

As early as the 1940s, in his book *What is Life?*, the physicist Erwin Schrödinger had wondered if biology could be reduced to molecules. It was reading this book that persuaded the young Englishman Francis Crick (1916–2004) to swap physics for a career in biology. Crick and the American biologist James Watson (b.1928) discovered the double-helical structure of DNA in the 1950s, setting in motion a molecular revolution of biology. Looking for the origins of life seems to take us, once again, on a journey down to the smallest structures.

The atom, for all its emptiness, is a sort of barrier between two worlds: the everyday world that appears to be made of separate moving objects and the strange world of quantum physics. The atom turns out to be a barrier that is difficult to breach, requiring the injection of huge amounts of energy. On the other side of the barrier is the world of particle physics. DNA is another useful conceptual resting point, separating the living from the inanimate. But when we investigate DNA we don't need to know about atoms or parts of atoms any more. The journey down to smaller structures is of no use to us now. DNA is a code that must be read to be understood, which makes it a different sort of barrier from that of the atom, and a different sort of structure from any we have encountered so far in the universe. With DNA, the universe takes a step into the symbolic.

The DNA code is written in just four chemical bases: adenine, guanine, cytosine and thymine. To make it even clearer that we are not concerned with the structure of these chemicals only that they make up an alphabet of life, they are usually reduced to their first letters: AGCT. Life is a code that can be translated and read as if we were reading a book. All of life (that we know of) is written in this code, another clue to its common origin. DNA is a very long molecule that can be read as a string of these four letters.

There are words in the language of life and they are each three letters long: ggg (guanine guanine guanine), ctg, atc, and so on. This ought to mean that the language contains 64 different words (4 × 4 × 4), but some of the words mean the same thing with the result that effectively the language contains only 20 words. In addition, three of the 'words' represent punctuation, which makes it a little different from languages we are more familiar with.

There are also sentences in the language of life. The sentences are what we call genes. Much of the long string of letters that is written on the long DNA molecule is what used to be called junk DNA, but embedded within the junk, or what is now more accurately called noncoding DNA are the sentences, or recipes, that are the genes. The genes are separated out from the non-coding DNA, as sentences are, by the use of punctuation. A gene is a sentence that makes sense, hidden among the non-coding DNA. The collection of genes and the non-coding DNA is called the genome. In some bacteria 90 per cent of the genome is gene sequences. In the fruit fly it is 20 per cent, and in humans it is less than 2 per cent. We are only just beginning to understand what the non-coding DNA does.

Some of it serves important functions and some really does seem to be junk. For example, many of the odour genes that are active in animals are inactive junk genes, in humans. These genes have become degraded in humans, becoming more and more useless, mutation by mutation.

Tiny machines called ribosomes find and read the gene sentences and make them into something physical. Word is made flesh. Each word in the sentence represents an amino acid. There are many different kinds of amino acid, but all we need to know is that life is constructed from just 20 of them (represented by the 20 different meanings of the words of the DNA code). For our purposes we don't need to know what amino acids are, only that they are a class of molecules with a particular structure. Not only is all of life written in a code that uses the same four letters, all life is written in the same language of 64 words. And these words always express the same 20 amino acids.

The sentence that is the gene is also a recipe that can be read and made into a string of amino acids. When the cook gets to the end of the sentence and has made the string of amino acids in the particular order specified by the order of the words in the sentence, the string folds up into a complex three-dimensional shape called a protein.[4] Every gene, then, is a recipe for a protein. The human body can make around 25,000 different kinds of protein, one for each gene.

Cells are factories full of tiny machines making proteins by reading selected recipes (genes) found on the DNA molecule. The cells of an organism serve different functions because the DNA tells the cell what functions it must perform. Cells express different proteins because different parts of our DNA molecule are switched on. A human cell might become a red blood cell because the protein that makes haemoglobin is switched

---

[4] The process is much more complicated than I make it out to be. For example, before the recipes can be read they are first transcribed on to another molecule, related to the DNA molecule and called RNA. But we don't really need to know this. I have concentrated on essentials.

on, and remains switched off in cells that become, say, brain cells. Haemoglobin is a complex protein (made by a complex protein) that has the ability to carry oxygen, in the blood stream, to the organs of the body. Cells in the liver produce a protein that breaks down food. Some cells manufacture keratin, a protein out of which nails and hair are made. Some hormones, the sex and adrenal hormones, for example, are proteins.

Proteins that switch the DNA on and off are attached to part of the non-coding DNA, but how this operation is performed isn't fully understood. The process of switching genes on and off involves a certain amount of circularity. How can DNA know how to switch itself on and off? It may be something that happens more at the cell level: cell signalling to cell to tell each other what they are, where they are in the body, and what genes should be switched on to produce the proteins needed to fulfil the cell's function. DNA seems to know what genes to switch on or off because of the context of the DNA within the chemical composition of a particular cell.

Organisms such as human beings are made out of many cells (there are also many organisms that are single-celled). Complex multicelled organisms like flies and human beings reproduce by a process we call sex. The parents donate half of the genetic information that will be carried forward to the next generation. The DNA molecule is so long that it is broken up into pieces called chromosomes on which genes are randomly distributed. Several genes that describe a single characteristic may be housed on different chromosomes. Every cell contains two sets of chromosomes (two sets of the DNA) except the egg and sperm cells which contain only one set each.[5] Sex allows one set from each parent to be recombined to make two new, and slightly different sets, in the next generation.[6] The small differences partly account for the generally small changes that we see between one generation of organisms and the next. In

---

[5] Exceptionally, red blood cells contain no genetic information.
[6] It has been calculated that there are around $2^{2,000}$ ways in which humans can be differently expressed. The biggest number in this book so far.

humans it might be a different eye colour. The shuffling of genetic information by reproduction results in more variation, which confers a selective advantage in a changing environment and allows greater complexity to evolve. A new organism is made from the repeated division and replication of a single cell: a single fertilised egg. Humans are made of some 100,000 billion cells. Duplication quickly leads to very large numbers as we saw when the universe inflated. Although the DNA in each cell is very close to being identical, cells do not all contain the same chemicals. Specific types of cell contain specific sorts of chemicals. During the process of division different sorts of cell emerge. In humans there are several hundred different kinds.

Before a cell divides the DNA in the cell is replicated. The structure of the DNA molecule has evolved so that it is peculiarly well suited to duplication. The famous double-helical strands of the DNA molecule always join together in the same way: adenine (A) always binds with thymine (T), and cytosine (C) always with guanine (G). These connections are called base pairs. In human DNA there are some 10 billion base pairs. If a strand of the double helix of DNA reads ATGGCGGAG, then we immediately know that that strand is attached to a corresponding strand on the other spiral that reads TACCGCCTC. Every time the DNA molecule replicates itself – when its 10 billion base pairs are copied – only about a dozen mistakes are made. A sophisticated copy-reading mechanism has evolved that ensures that errors are generally corrected. Each cell contains sort of proofreading enzymes. Even if a letter in the DNA sequence is changed it doesn't necessarily change the meaning of the word in the gene sentence (because a significant proportion of the words mean the same thing). But even when a change of letter does change the meaning of the word and a different amino acid is expressed, the overall function of protein produced often remains largely unaltered. Proteins can be made of a hundred or more amino acids, but often only a few amino acids control the function of the protein and the rest act as a sort of scaffolding. These copying errors are called mutations. Even when such a mutation does occur, it is even rarer that the mutation should happen in both pairs of genes. One non-mutant gene is often enough to ensure that the 'right' protein is expressed.

Usually, then, mutations result in small variations among organisms. The different blood types in humans are the result of tiny differences in the protein that controls the surface structure of red blood cells. A small change in the protein sequence of a particular protein involved in the production of melanin (the protein that determines skin colour) results in some people having red hair. Disastrous mutations are mostly avoided, but where they are not the organism expressed by those genes is quickly removed from the gene pool by natural selection. Mutations *are* the mostly small changes that Darwin argued for, and not the large changes that result from artificial selection.

*

The Human Genome Project was set up in 1990 with the goals of identifying all the genes of the human genome, and of reading the DNA sequence of the entire genome (to understand better the sequences that come between genes). At the time it was thought that there might be 100,000 genes, but when a complete map of the genes was published in 2003, it was discovered that there are, in fact, less than 25,000. Fruit flies and round worms have a genome at least half the size of ours, and rice has 40,000 genes. But even our relatively modest number of genes is enough to describe our complexity. When we speak of very large numbers we say that they are astronomical, but we might better name the largest numbers in the universe *biological*. The combinatorial possibilities of 25,000 genes is many, many orders of size larger than any numbers we have come across so far in our trawl across the universe. But even this is somewhat beside the point. The complexity of an organism is a consequence of the fact that it grows. It is the order and pattern of gene expression that accounts for much of the difference between species, not the relative number of genes in the genome. A gene called hoxc8 is switched on for longer in a chicken than in a mouse, which is what gives a chicken the longer neck. The genes that determine overall body plans are called homeotic genes. The plan of an eye is quite similar between organisms as different as humans and flies. In fact *all* eyes have expressed a gene called pax6. Jellyfish possess homeotic genes that are remarkably similar to the genes that determine our own body plan, and the similarity is best explained by evolution. We look different from jellyfish largely because the genes are expressed differently. It isn't vocabulary size that marks out a great playwright (though some do have large vocabularies); it is possible to write great plays with a very limited vocabulary. It's how the words are put together that matters, and that's definitely true of the language of life, which has a very small vocabulary and is written in few sentences.

*

Comparison of the genome (DNA molecule) across all living creatures tells us that all living things have a common origin. Evolution explains why the genetic code – the 64 triplets and the 20 amino acids that they encode – is the same across all living things. And evolution also explains how the messages written in that code have changed (sometimes hardly at all) over time and across species. Some genes are conserved over billions of years because any change would cripple some vital function, and such disadvantaged organisms would be removed by natural selection. How non-coding DNA changes over time, due to copying errors (mutation), is called DNA drift and can be used as a biological clock that times the appearance of common ancestors.

Given that proteins are made out of gene recipes, comparing proteins in different species is an equivalent way of showing that all species are related. There are proteins in humans and in yeast that fulfil very similar functions but are made of amino acids arranged in quite different ways. Comparison of such proteins in different species tells a story of how these proteins have evolved over time. The similar function of proteins in humans and in yeast points to the existence of a common ancestor of these two different living forms.

Evolutionists now have two techniques at their disposal for classifying living forms: the comparison of anatomical features of different fossilised and living organisms and the comparison of the DNA of living organisms.[7] Taken together these techniques turn evolutionary biology into a fully formed and testable science. Evidence of similarities at the DNA level can help confirm evolutionary counsinship at the macro level. In the late twentieth century molecular biology became a tool that can provide independent confirmation of fossil evidence. Darwin thought that dogs were descended from several different canines but recently, through genetic comparisons made at the DNA level, it has been discovered that dogs are all descended from the grey wolf. There was a time when the hippopotamus was thought to be closely related to the pig, but genetic evidence now tells us that it is more closely related to whales, dolphins and porpoises, the cetaceans. Sometimes organisms only look similar because they evolved under similar conditions. Morphology can be misleading. Earthworms and tapeworms may look similar, but they belong to different phyla (the largest grouping of organisms below kingdom). An analysis of DNA can tell a different story to that told by outward appearances. Anatomical comparisons once suggested that the gorilla is mankind's closest relative, but DNA analysis shows that we are slightly more closely related to the chimpanzee. Molecular biology revolutionised the way in which organisms are classified. DNA turns out to be another kind of fossil evidence.

The two dating techniques allow us to classify life into a hierarchy of connectedness back through chronological time. Making the journey in the other direction, it is tempting to

---

[7] DNA does not survive full fossilisation. From time to time some degraded DNA has been retrieved from animals found preserved in ice or mud. And see p. 250.

suppose that what we see is evolving complexity. It is hard to shake off the notion that bacteria are somehow more primitive than we are, but we are too wedded to an idea of what we think complexity should look like. We are biased towards complexity seen at the macroscopic level.

We may never find the common ancestor of all extant living forms: that part of the story of evolution is missing. But we do know that the common ancestor of all multicelled organisms was single-celled. We can find an array of single-celled forms in the deep past and our best current theory tells us that they must have a common ancestor even deeper into the past. Most of the branches appending single-celled life died out, but one branch leads to multicellular life and the rest descend to more evolved single-celled forms alive today.

Four billion years ago

At one time it was thought that all life was part of a chain requiring sunlight to thrive. Sunlight penetrates only the top 50 metres of the oceans, known as the photic zone, but in the 1970s single-celled life was discovered living near deep-sea hydrothermal vents, which are fissures in the earth's crust from where gas escapes, almost 2 kilometres deep. Instead of sunlight the organisms use chemicals in the water as a source of energy. Because of the pressure of the water, temperatures can reach hundreds of degrees Celsius at these vents. (The boiling point of water is 100°C at sea level and at current-day atmospheric pressure. Where the pressure is higher, so is the boiling point. At hydrothermal vents, the water pressure can be 25 times atmospheric pressure.) It is assumed that this is where the first life-forms lived, the only place they could thrive that was then protected from the damaging radiation coming from outer space. Single-celled life found by hydrothermal vents or in other extreme conditions are called archaea, a recent naming. They used to be lumped together with bacteria. It was once thought that archaea *only* lived in extreme conditions, but it is now known that such organisms live in a variety of conditions and today make up perhaps 20 per cent of the biomass of the planet.

Chemicals found in rocks 3.8 billion years old from Akila in Western Greenland may be the earliest evidence of life we have so far, but the findings are controversial. Other controversial physical evidence comes in the form of fossils of a similar date called stromatolites. They are to be found in Western Australia, and supposedly made of another early single-celled form called cyanobacteria,[8] basically pond scum.

As the seas cooled, life-forms evolved that use the energy of sunlight to turn carbon dioxide into sugar and oxygen, a process called photosynthesis. Cyanobacteria evolved either out of the kind of non-photosynthesising life found around deep-sea vents (archaea), or they both evolved separately from common ancestors lost to us. Cyanobacteria is the first upper-sea life and brings, as a waste product of photosynthesis, the first oxygen. There is evidence that cyanobacteria was dissociating carbon dioxide and releasing oxygen 3.5 billion years ago. All the world's oxygen is here as a gift of the photosynthesising bacteria kingdoms. Without bacteria there would be no oxygen.

The earth's first oxygen does not enter the atmosphere. Because it has never been in the presence of oxygen before, the earth must rust. Everything on earth that is capable of being oxidised is oxidised. We see evidence of this rusting in bands of red oxide deep underground. Only when this oxidation process has come to an end do the oxygen levels in the atmosphere begin to rise. This takes a long time. Even 2.4 billion years ago the oxygen level in the atmosphere had risen to only 0.1 per cent; 2 billion years ago it had reached 3 per cent. The current level is 20 per cent.

The first archaea protected themselves from the sterilising effect of ultraviolet radiation by living deep under water; the

---

[8] Or what used to be called blue-green algae. Nowadays the term algae is restricted to more complex forms of life.

first upper-sea organisms found protection by hiding behind grains of calcium carbonate, or chalk. Carbon dioxide dissolves in water as various carbonates. The first photosynthesising bacterial forms use calcium carbonate as a kind of shield. When the bacteria dies, the calcium carbonate grains sink and form what will eventually become layers of chalk. All chalk is made out of this living process, from what are effectively tiny skeletons too small to see with the naked eye. Limestone is another form of calcium carbonate, but since it is made out of the shells of more complex creatures it is not found on earth until billions of years later. This early use of calcium carbonate by bacteria is a clue to the relatedness of all living forms, evidence that evolution happens. The shields of primitive bacteria later become the protective shells of more developed organisms, and then bones – an ingenious adaptation of a shell worn on the inside and used to bear weight – of creatures that will eventually leave the buoyant world of the sea for land.

One of the effects of chalk formation is to lock away vast amounts of carbon and so help to stop the greenhouse effect from running away. Later, carbon will also be locked away in coal and oil, other products of living processes and geological ages of compression. Volcanic activity redistributes some of this carbon in a feedback loop that is evidence of what the English scientist and freethinker James Lovelock (b.1919) named Gaia. Lovelock first put forward his now widely accepted hypothesis in the 1960s, at which time it was ignored, becoming the subject of controversy in the 1970s. Gaia is a Greek personification of the earth, literally meaning grandmother of the earth. Gaia is the understanding that there is an intimate connection between the living and non-living processes of the earth. The oceans move heat around the earth and the mountains create weather systems as part of this balanced global feedback system, which extends out to include the activities of the moon and the sun. The greenhouse effect and the levels of oxygen in our environment have, over billions of years, become very finely tuned as a result of tectonic activity and living things. If it were not for the greenhouse effect, global temperatures would be 15 degrees lower than they are now, although these days we are more concerned that mankind will make, or already has made, the greenhouse temperature too high.

Archaea and bacteria developed sophisticated molecular machinery in the relatively short period (perhaps only a few hundred million years) after the earth became a suitable habitat for life. In the 1980s, Fred Hoyle and the Sri Lankan astronomer Chandra Wickramasinghe (b.1939) argued that it was not possible to construct such a complex machine, even the most basic bacterium, in the time available. They said that it would be as if a whirlwind could assemble a Boeing 747 from scraps in a junkyard. They calculated that the chance of assembling an amino acid by chance is 1 in 10^{20} and since a simple bacterium could be made from 2,000 proteins then the chance of assembling a bacterium from chance is at least 1 in $10^{20 \times 2,000}$ or 1 in $10^{40,000}$. Again, this shows how much larger biological numbers are than astronomical ones. The number of elementary particles in the visible universe is 'only' 10^{80}. But there is a flaw in Hoyle and Wickramasinghe's argument. Nature does not assemble complex things from scratch. Natural selection is the only explanation that is needed. So long as there is some small advantage, no matter how tiny, that advantage has an increased chance of being selected. If some protein caused a reaction to go a tiny bit faster, it has a better chance of being selected than other proteins that leave the reaction unchanged. (Just as important, any protein that made the reaction go that much slower would be taken out of the running.) Nature doesn't select from all possible configurations; it selects from what there is. It doesn't matter how primitive that protein is; all that matters is that the primitive something exists and works slightly better than anything else that is around.

How chemical life became biological life is still a missing part of the story of an evolving universe, but few scientists doubt that we will soon discover how inanimate atoms evolved into animated molecular structures. Life is a process of self-organisation and self-replication. Life is complex chains of chemicals called polymers that reproduce themselves. It is presumed that over a few hundreds of millions of years of evolution prebiotic molecules evolved into self-replicating molecules by natural selection. Quite how that happened is still not known, but few biologists doubt that it did happen.

The boundary between geological and organic life is blurry. Not only does life appear to have emerged from inorganic nature, but what we distinguish as life is inextricably bound up with the inorganic activities of the planet: the oceans, volcanoes, mountains and living processes are woven into a web of interconnection.

Newton thought that alchemy would reveal to him the divine something that makes stuff live, and that was also to be found in plants, animals and in certain inorganic forms like crystals. But when we know how life emerges from the inanimate, the organic and inorganic worlds will have become a continuous spectrum. Life will be an artificial distinction we make from the inanimate. We will be able to trace our evolution back not just to the common ancestors of archaea and bacteria but to primordial hydrogen, to the Big Bang, perhaps even to a multiverse or some such starting condition of our local universe. We will know how life is written into the laws of nature. We will be closer to answering the questions: What do we mean by life? And what other forms might life take? Our imaginations can take flight, to envisage other ways that life might have come to be, other ways that life could be.

Three billion years ago

A billion years after life first emerged on earth, it was entirely single-celled archaea and bacteria. In fact much of life today is still archaea and bacteria. The persistence of single-celled life-forms illustrates just how intimate the relationship between living and non-living processes can become. Archaea and bacteria are only the simplest life-forms in the sense that they are made from a single cell, though even a single-celled bacterium contains 10^9–10^{11} atoms. But having been around the longest, they are the most resilient forms of life and have some claim to being the most evolved.

Bacteria are essential to the continuing existence of life on the planet. They balance not just the levels of oxygen in the ecosystem, but nitrogen, carbon and sulphur. Bacteria are known to be involved in the formation of oil. Wood would not rot if it were not for bacteria. Some bacteria have been found in rocks 1,000 metres underground slowly digesting organic material without the aid of oxygen and dividing only once every thousand years or so, surely the most laid-back of all the earth's life-forms. Bacteria may even be involved in the depositing of metals underground.

Bacterial life forms did not become redundant when the higher life forms emerged, rather the opposite: they thrived and may be better adapted to survive than multicellular life forms. It is bacteria that are more likely to survive a future catastrophe. Bacteria have been found that can live in sulphuric acid, or even in nuclear waste. Crystals of magnetite found in some bacteria help them to orient themselves along the earth's magnetic field. Here is evidence that there are some forms of life that could thrive in places quite different from earth, evidence that so-called simple life does not in fact require the privilege of an earth precisely like this one.

It may be clear what sort of life-form is best suited to a future catastrophe. We would have to use our highly evolved and complex brains to outwit nature, but since it is out of nature that our complexity emerges surely that is a contest we cannot win. Bacteria, on the other hand, have long been tested against whatever nature has thrown at them, existing from the first in a world much more violent than the world of today.

Mankind has put itself in competition with the volcanoes in the carbon-balancing activities of the planet. When Gaia needs to rebalance the planet to take account of these and other disturbing activities, it is unlikely to be to mankind's benefit. If humans disappear, it seems likely that bacterial forms will remain and evolve further as the conditions allow.

Whether or not humans escape the wrath of Gaia, they will not manage to survive if bacteria disappear. We do not have an existence independent of bacteria. Not only are bacteria integrated into the life of the planet, they are integrated into workings of the human body. There are ten times as many bacterial cells in the human body as human cells, mostly on the skin and in the digestive tract.

Two billion years ago

Archaea and photosynthesising bacteria are single-celled organisms without a nucleus collectively called prokaryotes, literally meaning 'before a kernel': the Greek for kernel is *karyose*. Between 2 and 1.5 billion years ago the first single-celled eukaryotic life emerged, that is, cells that possess a nucleus. The nucleus houses and protects the DNA. It isn't known if eukaryotes evolved from prokaryotes but they must have had a common ancestor that was perhaps quite distinct from either. Modern DNA analysis suggests that there is no meaningful definition of a prokaryote; it is merely the collective name we give to two quite distinct ancestral lines called archaea and bacteria. It even begins to seem possible that archaea are more closely related to eukaryotes than to bacteria. Single-celled eukaryotic life appears to have emerged when oxygen levels became high enough (about 0.4 per cent) to support this more complex life.

What *is* widely accepted is a theory popularised by the American biologist Lynn Margulis (b.1938) in the late 1960s. The theory puts forward the idea that some prokaryotic bacteria inserted themselves into eukaryotic life as organelles, the collective name for two types of new structure: chloroplasts and mitochondria. Lynn Margulis suggests that, through symbiotic attachment, prokaryotes became the organelles of some other single-celled form. Evidence in support of such a theory comes from the fact that the DNA in organelles (which comes from the invading prokaryotes) is entirely distinct from the DNA found in the host nucleus.

Eukaryotes with chloroplasts are called protophytes and eukaryotes with mitochondria are called protozoa. A choloroplast is what in the future will make photosynthesis efficient in plants. In protozoa the mitochondria makes it possible for cells to use oxygen as a fuel for the first time. It is these organelles that will, when there are such things, separate the animal and plant kingdoms.

All life remains single-celled and confined to the oceans.

One billion years ago

It is thought that multicellular life first arose about 1.2 billion years ago. The landmass of the earth was a single continent called Rodinia.[9] Multicellular life presumably started out as another sort of symbiosis, loose cooperation between single cells that became more and more complex. Some photosynthesising eukaryotes developed in colonies as the first seaweed (part of a grouping called red algae). Sponges are another example of a primitive multicellular life form. A detailed mechanism for how the leap from single-celled to multicelled life was made is still missing. Some time afterwards, though again we do not know by what mechanism, some eukaryotes invented sex: the process in which an egg (a type of cell) is fertilised before division takes place.

Life may have been single-celled for 3 billion years, but when the mechanism is in place that allows multicellular life to emerge there is no reason why it should not emerge, in geological terms, almost overnight. In fact it must emerge as soon as the conditions make it possible.

Evolution appeared to speed up again about 550 million years ago with the first and sudden appearance of hard-bodied animals in the fossil record. The fact that there are no fossils earlier than this date had been a problem since before the time

[9] Rodinia had broken up again by about 750 million years ago.

of Darwin. He thought it was the single biggest threat to his theory of evolution. If it had been hard for him to account for the absence of intermediary forms between species, how much harder to account for the lack of any fossils at all older than those found in rocks half a billion years old.

Before the 1980s, before molecular biology had really got going, a meagre fossil record was all that was available to map out evolutionary development. But the fossil record was troubling: not only were there no fossils before 550 million years ago, an explosion of life-forms appears in one fell swoop. The evidence for the so-called Cambrian explosion is striking, particularly at the Burgess Shale, a fossil bed in the Canadian Rockies discovered in 1909, though the significance of the fossils found there only became apparent when they were re-examined in the 1980s. The find at Burgess Shale was brought to public attention by the American biologist Stephen Jay Gould (1941–2002) in his book *Wonderful Life*, published in 1989. It became clear that there are long periods of geological time when nothing much seems to change, followed by rapid periods of evolutionary development.

In 1954, the American biologist Ernst Mayr (1904–2005) had observed that large populations remain stable: they do not exhibit the kind of evolutionary changes that the theory of natural selection appears to predict. What this evidence of evolutionary stasis means was first discussed by Gould and the American biologist Niles Eldredge (b.1943) in the 1970s. The evidence of the Burgess Shale find appears to be further evidence that stability is the norm followed by sudden periods of change, what Eldredge and Gould called punctuated equilibria. How to explain why populations remain stable for millions of years even though there is change at the genetic level became a subject of spirited, sometimes acrimonious, debate.

Evolution can be reduced to competition between genes, an understanding made popular by the British evolutionary biologist Richard Dawkins (b.1941) in his famous book *The Selfish Gene* (1976). Evolution has to happen at the genetic level because genes are the only way that a trait can be passed on from generation to generation. But natural selection doesn't only work on genes, it works at every order of size. Eldredge says that Dawkins's view works best when evolution is being investigated from generation to generation but 'is useless as a general evolutionary theory covering the large-scale events in the history of life'. Relating evolutionary changes at the organism (or at the population) level to changes at the genetic level is very hard to do. What is missing, and is still argued over, is a mechanism to explain the interaction between the environment, genes, and organisms expressed by genes.

It makes little sense to ask which is the more fundamental: the gene or the population. It's like asking which is the more fundamental: the chair or the atoms it is made from. If we are talking about chairs then a chair has no meaning at the atomic level, since atoms possess no properties of chairness. Large structures (chairs and cats) possess qualities (chairness and catness) that only exist in the macroscopic world, that are not qualities of the parts from which they are made. But if we want to know what a chair is made out of, then we must ultimately talk about atoms and subatomic particles.

The Cambrian explosion may only appear to be an explosion because it emerges at a scale of size that we humans visually acknowledge as complex. We grant ourselves a privileged position if we think that these new forms are any more complex than the forms that came before, whose complexity may only be apparent at molecular levels. We are in danger of being anti-Copernican if we confer specialness on something just because it happens to be the kind of size that we are. The change that brought multicellular life-forms into existence for the first time is such evidence that evolution appears to speed up. And the first appearance of hard-bodied organisms is more evidence. But whether or not evolution actually speeds up or merely appears to is, perhaps, a matter of emphasis. It appears to speed up because we humans take more notice of multi-celled organisms, or hard-bodied ones.

Because of its historical and visual significance, the Cambrian period has become the hub around which geological dating takes place. The half-billion year or so swathe of time since the Cambrian explosion is called the Phanerozoic eon and runs to the present day. Phanerozoic literally means 'visible life'. All of geological time that came before the Cambrian explosion is called the Precambrian eon, a disproportionately long run of time that stretches back to when the earth was formed 4,500 million years ago.

Eons are further divided into eras. The last era of the Precambrian eon is named the Neoproterozoic era and runs from one billion years ago to the beginning of the Phanerozoic eon 542 million years ago, which is also the beginning of the Palaeozoic era and the beginning of (yet another subdivision) the Cambrian *period*. Until recently the evidence of large-scale independent life forms began with the Cambrian, but during the twentieth century evidence began to be collected of large-scale organisms living in the period preceding the Cambrian, though it took some time to interpret the finding. This period, called the Ediacaran period, was named in 2004. We join named geological time at this point.

The end of the Precambrian eon

Neoproterozoic era (1 billion to 542 million years ago)

Ediacaran period (630 to 542 million years ago)

Although there is no evidence of large-scale independent life forms from before this period this does not mean that there aren't any. The first macroscopic creatures that we know of are globs of matter called Ediacaran biota that look like 'mud-filled bags, or quilted mattresses'. Ediacaran life appeared in shallow reed beds that allowed the large surface area of these organisms to absorb as much oxygen as possible. Some of these life-forms are the first mud-burrowing creatures, a skill acquired via the evolution of a coelom, a sac that contains the organs. It not only allows the bodies of early creatures to tunnel, it will later allow larger organisms with a skeleton to bend and twist. There is no complex life on dry land. Ediacaran biota seem to be unrelated to any forms that came later, which must mean that Ediacaran biota and other organisms are already separated by long lineages and must have common ancestors in the deeper past that are either not preserved in the fossil record, possibly because those earlier ancestors were too soft to fossilize, or made fossils not yet discovered.

Creatures similar to jellyfish may have emerged at this time. (It was, until recently, thought that they emerged later.) Animals related to jellyfish belong to a phylum called Cnidaria. Certain symmetries in the body plan of jellyfish relate them to what we take to be the more complex life that followed.

A jellyfish has no respiratory, circulatory or excretory organs, no nervous system, and no obvious head except for where there is a mouth, and yet they do have discrete organs. Jellyfish don't have brains but they have several eyes. They are very long-sighted, but their vision is good enough to tell night from day and up from down.

It seems that from the earliest times the eye and brain developed together. Our two ways of seeing, with the eyes (where light is received) and with the brain (where light is interpreted and understood), are connected. There are clumps of molecules within the cells of many organisms, including many single-celled life-forms, that are made from a light-sensitive protein called rhodopsin. Once this protein evolved it subsequently appears in all eyes for all time. Though it is not yet understood how the rhodopsins in prokaryotes are related to the rhodopsins found in evolved eukaryotes, it seems likely that they are related, and that this family of proteins will turn out to be further evidence of evolution from common forms.

The earth had become a single landmass again, called Pannotia, about 600 million years ago. It was not long lasting, breaking up after only 60 million years.

The beginning of the Phanerozoic eon
Palaeozoic era (542 to 251 million years ago)

Cambrian period (542 to 488 million years ago)

The dating of the Cambrian period has changed significantly even in the twenty-first century. Some authorities have the Cambrian beginning 570 million years ago, or earlier, but in 2002 the International Subcommission on Global Stratigraphy, an internationally regarded body charged with defining such parameters, put the beginning of the Cambrian at around 545 million years ago, a date that was revised to 542 million years ago in 2004. Rocks from this period were first discovered in Wales. Cambrian is taken from the name of an ancient Welsh tribe.

The first shelled animals appear in the Cambrian period. They are a particular order of shelled animals called arthropods, distinguished by their hard exoskeletons and segmented bodies. The body plans of all animals alive today are inherited from animals that first appeared in the Cambrian period. Trilobites – strange creatures that look like very large woodlice – are the first arthropods. Surviving for over 200 million years, they are probably the second most famous fossils after the dinosaurs. At one time, they swarmed the seas as one of the most prolific, of all life forms. Arthropoda is the largest phylum of non-extinct animals and includes insects, spiders and crustaceans (all of which appeared much later). Many phyla appear in the Cambrian that are now extinct. It is possible that some animals make the first tracks across land during this period; otherwise all animal life is confined to the oceans. There are no land plants. The surface of the earth is made up of barren deserts and oceans.

During the Cambrian the landmass of the earth is arranged as two major continents, called Gondwana and Laurentia.

At the end of the Cambrian, a period of extinction is thought to have occurred during which many species are eliminated or severely reduced.

The divisions between geological *periods* of time are marked by massive extinctions, which result in a distinct change in the fossil record. The reasons for the extinctions are largely conjectural but may mostly have come about as the result of increased volcanic activity and subsequent changes in greenhouse effects. It takes a cataclysm on this scale to upset the strong stability of populations. Any less dramatic change to an ecological system does not propel evolution. Fires and storm cause only temporary disruption. Ecosystems can reassemble themselves out of the information contained in nearby populations, where the evolutionary history is little different from the local population that has been destroyed. A population will easily adjust to a slow-moving glacier so long as new breeding grounds can be found to preserve the status quo. It requires disruption so severe that the whole planet is involved for rapid evolution of new forms to take place: an earthquake, meteorite impact or sudden change in the amounts of greenhouse gases in the atmosphere. It is this general stability at the population level that explains why evolution appears to progress rapidly when cataclysmic changes occur to the planet.

Ordovician period (488 to 444 million years ago)

Early geology was dominated by British geologists. This period too is named after an ancient British tribe, as is the next period. There are fossils of liverworts – something between a moss and seaweed – that are 475 million years old. The first primitive fish appear. They are the first vertebrates (animals with a backbone). Between the Ordovician and Silurian periods the first of the so-called five major extinction events occurs, during which half of all existing fauna became extinct. The reason for the extinction is not agreed on, though it may have coincided with the onset of a particularly severe ice age.

Silurian period (444 to 416 million years ago)

Around 440 million years ago the first air-breathing animals begin to colonise land. They are tiny mite-like creatures. At the same time, the first true plants also begin to colonise land, though there are more controversial accounts that suggest there were land plants from much earlier times. The first land animals begin to break down the first vegetable matter, to make the first compost.

Fossils of macroscopic land plants have been found in Ireland that are 425 million years old.

During this period most of the landmass of the earth was in the Southern hemisphere. What is currently the Sahara desert was at the South Pole.

Devonian period (416 to 359 million years ago)

About 400 million years ago there are ferns, millipedes and spiders, and primitive sharks. There is evidence of the first freshwater life. Jawed fish with skeletons proliferate. The coelacanth is such a fish. It emerges about 390 million years ago, and outwardly remains unchanged to the present today. It was once thought to have become extinct more than 60 million years ago, until a specimen was caught off the coast of South Africa in 1938. Seeds appear for the first time. The biomass shifts from the sea to the land. The soil is full of mites and millipedes and there are carnivores to eat them, the first time that animals begin to eat each other. Amphibians emerge from fishes similar to the coelacanth. By the end of the period there are salamander-like amphibians among the first of the land vertebrates. At the end of the Devonian the second of the five major extinctions occurs. Recent thinking has it that the extinction happened in pulses over a 3 million year period, and for various reasons. By the end of the Devonian a third of all families of living things become extinct.

Carboniferous period (359 to 299 million years ago)

Oxygen levels rose to a peak of 35 per cent. Massive fern-like trees, unrelated to modern trees, appear. They will fossilize as coal. There are millipedes up to 2 metres long. The oldest physical evidence of animal life, other than fossilized bones, is a fossilized millipede track made 350 million years ago and found in Scotland. The first reptiles appear, evolving out of a common ancestor with amphibians. By the end of this period reptiles have become fully independent of water. Arthropods begin to leave the sea.

Permian period (299 to 251 million years ago)

Some insects (arthropods) evolve the ability to fly. Dragon-flies are one of the first creatures to conquer a new element. At the end of the Permian the largest extinction occurs, destroying up to 96 per cent of all species. The earth is a single landmass at this time, called Pangaea, and largely desert. The Great Dying lasts for 60 million years. There is some tentative evidence that a meteor may have been the cause.

Mesozoic era (251 to 66 million years ago)

Triassic period (251 to 200 million years ago)

The first trees (conifers) appear. Some lizards evolve into crocodiles and others into dinosaurs. There are bees. There is another major extinction at the end of this period.

Jurassic period (200 to 146 million years ago)

Around 200 million years ago Pangaea begins to break up. The Jurassic period is dominated by reptiles, particularly dinosaurs. The Archaeopteryx and flying reptiles evolve. Around 200 million years ago turtles emerge and also rodents. There are flies, termites, crabs and lobsters.

Cretaceous period (146 to 66 million years ago)

Some of the first mammals are marsupials. There are primitive kangaroos 136 million years ago. It was once thought that mammals only appeared after the disappearance of the dinosaurs, but they overlapped for at least 65 million years; they just did not flourish. All the major groups of insects now exist. Flowering plants are latecomers, arriving about 75 million years ago. Avian dinosaurs evolve into birds.

The last major extinction, apart from the one that is currently happening, occurred 65 million years ago when the dinosaurs became extinct. Though the reasons for most of the major extinctions are conjectural, the reason for this extinction is widely agreed on. In 1978 the American geologist Walter Alvarez (b.1940) and his physicist father Luis Alvarez (1911–1988) put forward the idea that a comet 10 kilometres across hit the Earth at around this time. In May 1984 evidence was discovered of a comet 30 kilometres across that hit the Yucatan peninsula in Mexico 64.4 million years ago, the largest collision in the solar system since the end of the late bombardment. There are still some supporters of the theory that it was a volcano that did for the dinosaurs. Global catastrophe interrupts periods of evolutionary stasis and clears a niche for rapid evolution to occur. The removal of the dinosaurs, as well as half of all other animal species, allows the mammals to flourish, though most of the mammals that evolved early on later die off.

Cenozoic era (66 million years ago to the present day)

Palaeocene period (66 to 23 million years ago)

It is not certain when the first mammals with placentas appeared, but fossil evidence tells us that they were definitely around by the early Palaeocene period. DNA evidence would suggest that there was a common ancestor of all placental orders between 100 and 85 million years ago. Without fossil evidence this prediction remains controversial. The first primates appear in the early Palaeocene period. Again, the DNA molecular clock suggests that there was a common ancestor much earlier, probably in the mid-Cretaceous.

Rabbits and hares appear 55 million years ago. The Himalayas begin to rise 50 million years ago. The face of the earth looks recognisably as it is now, except that Australasia is attached to

Antarctica. Bats, mice, squirrels and many aquatic birds (including herons and storks) appear during this period, as do shrews, whales and modern fish. All major plants make their appearance and grasses evolve.

About 30 or more million years ago ice ages become a regular feature of life on earth. Earlier ice ages had been dramatic but sporadic. An ice age is a period when there are ice sheets in the northern and southern hemispheres, which means that technically we are still in one.

By 26 million years ago, there are prairies widespread across North America. With the arrival of grasslands, animals evolve that graze on it, for example horses. Primitive apes also inhabit the grassland; and there are pigs, deer, camels and elephants.

Neogene period (23 million years ago to the present day)

Miocene epoch (23 to 5.3 million years ago)

Periods are further divided into epochs. Now that we are getting closer to our own time we can look into geological time more closely. All modern bird families are present. Many mammals evolve into recognisably modern genera: for whales it is the genera of sperm whales. Brown algae, also called kelp, emerges, allowing new species of sea creatures, such as otters, to evolve.

There are about 100 species of ape living during this epoch. Molecular evidence suggests that chimpanzees, gorillas and hominids begin to diverge somewhere between 15 and 12 million years ago.

The Mediterranean repeatedly dries up over a period of 1.5 million years: about 40 times from around 6 million years ago. This kind of detail is not available to us from the deeper past, a reason why our story only appears to get more complex as we get closer to our own time.

Pliocene epoch (5.3 to 1.8 million years ago)

At the beginning of the Pliocene the common ancestor of humans and chimpanzees moves about on the open savannah.

Our bipedal ancestors began to emerge between 5 and 3 million years ago.

Two or so million years ago marks the beginning of the most recent ice-age. Interglacial periods last between 60 and 100 thousand years and take 10,000 years to subside. (Again, this sort of detail is not available to us in the deeper past.) At the height of this ice age, ice covers three times as much of the globe as it currently does. The sea level drops by up to 130 metres. The interglacial periods of this current ice age could be an effect of the Himalayas acting as a barrier to atmospheric circulation. Alternatively, it is conjectured that these interglacial periods are mainly the result of slight changes in the earth's orbit and axis rather than, as it was in the deep past, the changing temperature of the sun or of changing greenhouse effects.

Pleistocene epoch (1.8 million to 11,800 years ago)

Great ice sheets repeatedly advance and retreat across North America and Eurasia, in places extending down to the 40th parallel, to Denver or Madrid. The temperature varies wildly during the last 900,000 years. It is always colder than it is now, sometimes much colder. The grasslands retreat and there are cold, dry deserts. The most recent glacial period began about 70,000 years ago, reaching a peak of coldness 21,000 years ago.

Holocene epoch (11,800 years ago to the present)

At the beginning of this epoch the weather gets warmer. Mankind begins to farm for the first time.

We are still coming out of the last glacial period, which partly explains why the north polar ice-cap, ice in Greenland and Alpine glaciers are melting. But there is widespread agreement that they are melting much faster than they should because of mankind's influence.

Today, there are 1.8 million species known to science. There are millions of species of microorganism still unnamed. The species that are alive today are an unimaginably tiny fraction of all those that ever lived.

*

Mankind has emerged by chance out of the vastness of deep time and the multiplicity of evolving life-forms. In this sense, the story of the unfolding complexity of the universe is clearly not about us. So have we lost our way among the diversity of nature? Do we find ourselves addressless and without privilege? Not entirely, or not yet, at least. We are the first species we know of that has the power to describe the world of which we are a part. That would appear to make us very privileged indeed.

In and Out of Africa

I knew at a glance that what lay in my hands was no
ordinary anthropoidal brain. Here in lime-consolidated
sand was the replica of a brain three times as large as
that of a baboon and considerably larger than that of
any adult chimpanzee.[1]

> Raymond Dart in 1925, on first holding the skull of
> what would later be named *Australopithecus africanus*.

Is there any reason to suppose that a story of increasing
complexity turns our attention to humans? Why not to other
living forms – bacteria, for example – and why even living
forms? We feel instinctively that the universe has evolved ever
more complex structures, but proving it is not easy. As the
physicist Eric Chasson has observed, 'The most primitive weed
... is surely more complex than the most intricate nebula in
the Milky Way.' But even if we suspect this to be true it is hard
to say why. Chasson has come up with a way of organising a
hierarchy of complexity based on how much energy is processed
by any given system relative to its size, be it a weed or a galaxy.

---

[1] Raymond Dart, *Adventures with the Missing Link* (1959).

As we might suspect, weeds process more energy relative to their size than galaxies do. Such strategies make it possible to believe that brains may indeed be the most complex structures in the universe, at least that we know of. There are even reasons to suppose that the human brain – a network of some 100 billion neurons in which each neuron is connected to as many as 10,000 other neurons – is the largest or close to the largest brain of any living organism compared to body size, though strictly by this reckoning the laurel may have to go to the shrew.

The fact that there is this story at all is because we humans are telling it. We have to think in order to describe the world, and thoughts belong to the mind, which as materialists we believe to be an emergent property of the brain. If the brain is an inextricable part of the story of unfolding complexity then we may feel justified in moving the story to that of increasing brain size in our human-like ancestors. As Copernicans we will want to believe that there are many other story-tellers out there telling a similar story.

*

All that remains of our human-like ancestors is a modest array of fossilised bones. Most of these few thousand specimens are shards, only very rarely a whole skull or a complete skeleton. And of these few thousand, it is out of a mere few hundred that palaeoanthropologists[2] have constructed a story of human descent. It is precisely, and ironically, because there is so little physical evidence to go on that ingenuity and imagination, together with a huge dose of speculation, propel research in this relatively modern field of scientific investigation.[3]

---

[2] From the Greek words *palaios* 'old' and *anthropos* 'man'.
[3] Curators of saintly remains, if such employment exists, must be faced with a similar task: how to separate from among the mostly bony relics of the past the genuine from the spurious. Roman Catholic relics are divided into three classes. First-class remains are the body parts of a saint or anything directly associated with the life of Christ (the crib or the cross, for example).

Palaeoanthropologists are particularly interested not just in fossils of skulls or parts of skulls but of the pelvis and parts of the pelvis. The story they tell is of increasing brain size and of bipedalism.

The gappiness of the fossil record is even more apparent and debilitating now that our focus narrows to the few tens of millions of years that trace the development of humans from out of the first primates. In the last few decades, fossil evidence has been supplemented by DNA evidence that is beginning to help fill in some of the gaps. DNA evidence is not usually retrievable from fossils,[4] and so for the moment we know more about our relatedness to other living species[5] than we do to the many human-like species that either survive only as fossilised bones or do not survive at all. But this is changing. In 2006, in north-eastern Spain, the first fossilised

John Calvin once said that if everything that claimed to be part of the True Cross was brought together, there would be enough material to build a ship. However, a study made in 1870 revealed that all such relics combined would weigh less than 1.7 kilograms. A second-class relic is anything with which a saint came into contact during his or her lifetime (an article of clothing, for example), and third class any object that has come into contact with a first-class relic, perhaps a cloth that once wrapped a saint's dead body. St Peter's in Rome is home to four major relics, though the Church makes no claim for them, and the same relics are to be found elsewhere: there is part of the True Cross, the Holy spear that pierced the side of Christ, St Andrew's head and Veronica's Veil, a cloth that bears the image of Christ's face. Scattered across the world there are three heads of St John the Baptist, two bodies of Pope Sylvester, 28 thumbs and fingers of Saint Dominic, which creates a curious problem that the taxonomist of this specialised field has to address. I once visited a museum in Siena where many significant saintly remains were carefully attributed, but more striking to me were the jars of bones at the furthest reaches of the gallery that were simply labelled 'Varie santi', as if in despair at any more definite attribution.

[4] The plot of Michael Crichton's novel *Jurassic Park* (1990) is predicated on the retrieval of dinosaur DNA not from fossils but from blood sucked up by a Jurassic mosquito that has been found preserved in amber.

[5] That relatedness can itself be used to trace human evolution. Humans have co-evolved with microbes inside them. We can, for example, trace human ancestry back to Africa through a mouth bacteria called *Streptococcus mutans* (*New Scientist*, 18 August 2007).

bone marrow was discovered, preserved in the fossilised bones of 10 million-year-old frogs and salamanders. Bones are not always fully petrified in the fossilisation process and in some very rare instances, as in this instance, soft tissue can be preserved within the partly petrified bone. It is conceivable that such finds could be made by re-examining current collections, though curators would presumably be reluctant to break open their rare artefacts on the off chance of making an even rarer discovery.

*

Using the various techniques currently available, the story goes that sometime between 100 and 65 million years ago a lemur-like primate evolved out of insectivore ancestors. About 25 million years ago this lemur ancestor of ours began to diversify into the simian primates: Old World monkeys, New World monkeys, and apes. Humans ultimately descend from the latter branch: the apes.

Some 19 million years ago the apes had diverged into the lesser and greater apes. The gibbon, for example, is a lesser ape. Any lesser or great ape and all their extinct relatives are called hominoids. There were dozens of hominoids at one time. The hominoids further diversified and one branch is the African and the Asian great apes and their extinct relatives, collectively called hominids. The orang-utan is the sole survivor of the Asian great apes. Molecular evidence suggests that the orang-utan diversified from a common ancestor with the African apes between 12 and 16 million years ago. Fossil evidence cannot help us here since the fossil record disappears from Africa about 16 million years ago, only to be picked up again 5 or 6 million years ago. In between times, fossils of great apes appear in Europe and Asia. It is assumed that the great apes continued to evolve in Africa and that perhaps the acid soil of the rainforests made an unsuitable environment for fossil making. A considerably less popular but alternative proposition is that the great apes left Africa to evolve further in Europe and Asia before returning to Africa, where the story continued. Whatever happened, all hominid fossils dating from the last several million years have been found in Africa, pointing to our origin there, as Darwin had suspected.

The African great apes descended to just three survivors: chimpanzees,[6] gorillas and humans, a group (including their many extinct relatives) called hominines. There are two surviving species of chimpanzee: *Pan troglodytes*, the common Chimpanzee, and *Pan paniscus*, known as the Bonobo (or archaically as Pygmy Chimpanzee). There is, unfortunately, no evidence of chimpanzees in the fossil record.

Molecular biology tells us that humans and gorillas diverged between 8 and 6 million years ago, and humans and chimpanzees perhaps 5 million years ago. [7] The many extinct bipedal ancestors that diverged from ancestors of gorillas and chimpanzees are called hominins.

Humans are the sole survivors of the genus *Homo*. They are also the sole survivors of the species *Homo sapiens*, and the only members of the subspecies *Homo sapiens sapiens*. In some accounts Neanderthals are the subspecies *Homo sapiens neanderthalensis*, but it is perhaps even more convincingly argued that they are a separate species, *Homo neanderthalensis*.

To break apart these Russian dolls the palaeoanthropologist must somehow decide whether a fossil looks more ape-like (by which is meant more like a chimpanzee or gorilla) than human-like, and sometimes it seems as if these distinctions are somewhat arbitrarily made. The similarity between man and ape had presumably always been apparent whenever man and ape met. When Queen Victoria visited Jenny, an orangutan on display in London in 1839, she noted in her diary that the creature was 'frightful, and painfully and disagreeably human'. Maybe Jenny was just as astonished. The queen

---

[6] Confusingly, chimpanzees are sometimes casually and popularly called monkeys; but to be precise, monkeys are not even apes.

[7] There is no agreement about when humans and chimpanzees diverged. It might well have been as long ago as 8 million years.

acknowledged a fear that even the pre-Darwinian idea that species are fixed and separately created could not allay. For a queen, whose very existence is dependent on a belief in the superiority of breeding, the closeness that repulsed her must have been all the more unnerving. Darwin visited Jenny too, and had a different experience, writing in his diary in unabashedly anthropomorphic terms that she was full of the joy of a naughty child. 'Man in his arrogance thinks himself a great work ... More humble and I believe true to consider him created from animals.' This response to the visceral impact of being in the presence of a distant cousin was recorded 20 years before the publication of his theory of evolution.

The oldest hominin ancestors cannot be categorically claimed even among the few known contenders. *Sahelanthropus tchadensis* fossils are dated to 7 million years ago. Some authorities cite it as the oldest known ancestor of the genus *Homo*, but that is based on the assumption that chimpanzees and humans diverged longer than 7 million years ago, rather than 5 million years ago as molecular analysis might suggest. If the more recent date is taken, then *Sahelanthropus tchadensis* might well be on the branch that descended to chimpanzees rather than humans. Similarly, a fossil of *Ardipithecus ramidus kadabba* dated between 5.8 and 5.2 million years ago is probably on the chimpanzee branch rather than the human one. A case is also made for *Orrorin tugenensis*, another example of a hominin ancestor, again most likely on the chimpanzee or gorilla branch. *Orrorin tugenensis* lived between 6.1 and 5.8 million years ago.

There is more general agreement that humans are on some branch that descended from a genus of hominins called *Australopithecus*. Australopithecines first appeared about 4 million years ago.

But there are two kinds of *Australopithecus*: the relatively slender (anthropologists use the word 'gracile') ones that were, perhaps, closely related to whoever our direct ancestors were, and the more robust ones, the paranthropines (members of the genus *Paranthropus*), with their big plant-chewing molars, who split off from the gracile ones around 2.7 million years ago and died out a little over a million years later, long after early humans had emerged in parallel.

A genus of Australopithecines named *Australopithecus afarensis* seems to have lived between 4 and 3 million years ago and evolved into various other hominins. The most famous *A. afarensis* is Lucy, also known as AL288-1. She was discovered in 1974 in Ethiopia and her fossilized bones are 3.18 million years old. Her brain capacity was about 380 to 430 millilitres, between a quarter and a third the size of a human brain.

Australopithecus africanus was about 1.2metres (4 feet) high, and on the evidence of pelvic bones and teeth seems to have been more human-like than ape-like, with a brain about the size of a chimpanzee, and larger, at 485 millilitres, than the brain of *A. afarensis*. The lineage of *A. africanus* is unknown, nor is it known what it evolved into, except that it was around just as *A. afarensis* was disappearing, and disappeared itself 2.5 million years ago.

Between 2.5 and 1.5 million years ago there were at least five species of *Australopithecus* and *Paranthropus* coexisting in Africa, but none is thought to have played a direct part in the evolution of the *Homo* genus. Humans are thought not to have descended directly from any of these extinct hominins. They were all transitional forms, and to make any greater claim than that is too speculative: there are too many gaps in the fossil record. Distant cousinship, not direct descent, is as much as we can hope to trace. All we can say is that about 2 million years ago some genus of hominin, whose precise existence will almost certainly remain hidden from us forever, evolved into the genus we call *Homo*.

The earliest known species in the *Homo* genus is arguably *Homo habilis*, named by the Kenyan Archaeologist Louis Leakey (1903–1972) in 1964. Its brain size was on average between 590 and 650 millilitres, somewhat larger than the brains of australopithecines. It is thought to have emerged about 2.2 million years ago. Stone tools have been found around remains of *Homo habilis* (though the use of stone tools is not uniquely characteristic of the *Homo* genus, and pre-dates even *Homo habilis* by at least 0.3 million years). In other respects *Homo habilis* is the least human-like of all the ancient Homo species, and authorities such as the Kenyan palaeontologist Richard Leakey (b.1944) exclude *Homo habilis* (because of its small size, disproportionately long arms, and other non-human-like characteristics) from the *Homo* genus. For them it is instead named *Australopithecus habilis*.

To add to the confusion, *Homo rudolfensis* may be an earlier species of the *Homo* genus (though the attribution is highly contested) and from which *Homo habilis* is descended. *Homo rudolfensis* is named on the basis of a single skull and was at first thought to be an example of *Homo habilis*. Its relatively large brain capacity of 752 millilitres was re-estimated as 526 millilitres by the anthropologist Timothy Bromage in 2007.

The paucity of the fossil record illustrates the fragility of constructing anything in this field that could be counted as scientific proof. This is not bad science so much as a branch of science struggling to make the best of what little hard evidence there is. It has been said, perhaps apocryphally, that some bone fragments have been arranged in every possible permutation of descent in order to bolster different theories. The palaeoanthropologist John Reader has wryly observed: 'It is remarkable how often the first interpretations of new evidence have confirmed the preconceptions of its discoverer.' Which might be taken as evidence in support of Nietzsche's observation that every theory is a private confession. Until very recently fossils and tools were the only way we had of dating our evolutionary past, but now DNA analysis begins to help harden scientific descriptions of the archaeological past.

Homo ergaster emerged around 1.9 million years ago. There is general agreement that here at last is a species that definitely belonged to the genus *Homo*. It had a brain of around 1000 millilitres, over twice the size of *A. africanus*, and looked rather like we do. This does not mean that *Homo ergaster* was a direct ancestor, but what we do know is that in the space of a million years or less the brain size of some hominins had at least doubled. *Homo ergaster* had disappeared, perhaps grading into the species *Homo erectus*, by 1.4 million years ago.

The skeleton of Turkana Boy, discovered at lake Turkana in Kenya in 1984, is dated at 1.6 million years old and is classified sometimes as *Homo ergaster* and sometimes as *Homo erectus*. The skeleton is of a boy aged between 11 or 12 who was 1.6 metres (5 feet 3 inches) tall when he died. It is thought that had he lived longer he might have grown to 1.85 metres (6 feet 1 inch).

Homo erectus overlapped and outlived *Homo ergaster*, dominating the earth for a million years. (*Homo erectus* may be what *Homo ergaster* evolved into.) About 1.5 million years ago *Homo erectus* leaves Africa, perhaps the first species of the genus *Homo* to migrate out of Africa and spread across the whole world. Famous fossilized specimens of *Homo erectus* include Java Man found in Indonesia, and Peking Man found in China. Since the 1990s, however, even this theory has been thrown into doubt.

Early fossils of species belonging to the genus *Homo* are beginning to be found across Eurasia: in Indonesia, Georgia, and Spain, for example. Some of these fossils appear to pre-date those of *Homo erectus*. If this part of the Africa story is to hold, then either there were African species belonging to the genus *Homo* that left earlier than was previously thought or we need a new story. For many scientists, Eurasia is replacing Africa as the new hot spot where early human evolution flourished.

However, around 350,000 years ago, the story is taken back to Africa, where a new species belonging to the genus *Homo* emerges called *Homo heidelbergensis*. Some time later, part of the population migrates and evolves into *Homo neanderthalensis*, popularly known as the Neanderthals. Some authorities name the Neanderthals *Homo sapiens neanderthalensis*[8] but in 1997 evidence suggested that the Neanderthals were genetically quite

---

[8] Modern man is more precisely designated *Homo sapiens sapiens* for this reason, but we do not know for sure whether there were any other subspecies within this species.

distinct from humans so on this basis do not appear to be of the same species nor direct ancestors.[9]

Between 150,000 and 200,000 years ago the story returns to Africa yet again, where a population of some unknown species in the *Homo* genus evolves into *Homo sapiens*. The oldest fossilized remains of *Homo sapiens* are thought to be between 130,000 and 195,000 years old, and are called the Omo remains after the river Omo in Ethiopia where they were found. There are several finds of human remains in the Middle East that date from 100,000 years ago, but these populations seem to have either died out or returned to Africa. The next-oldest fossil remains of a modern human were found in Mungo, Australia, and date from just 42,000 years ago.

Palaeoanthropologists tell us that mankind emerged out of Africa, and certainly all the oldest fossilized specimens of modern humans have been found in Africa. Molecular biologists help to confirm this account. In each cell we carry two kinds of DNA. There is the DNA contained in the nucleus and there is a distinct DNA called mitochondrial DNA, which lies outside the nucleus. Crucially, mitochondrial DNA is unchanged except by genetic mutation from generation to generation. Unlike DNA from the nucleus, which is divided into two from one generation to the next, mitochondrial DNA is almost entirely inherited from the mother. By calculating how much mutational drift has occurred across the mitochondrial DNA of the world's population, it has been possible, by calculating which parts of the world have inherited which mutations in their mitochondrial DNA, to divide the world's population into a series of mother clans. Just as siblings who share a mother belong to the clan of that mother, so we can connect all cousins that share a grandmother and all those descendants that share a great-grandmother, and so on. The world's 6.5 billion inhabitants can be arranged into just 33 mother clans of which 13

---

[9] Some accounts argue that *Homo erectus*, Neanderthals and *Homo sapiens* interbred in complex ways and that there was no uniform replacement of the older types by anatomically modern humans.

are in Africa. And these clans, too, converge on a single clan and a single mother living in Africa about 150,000 years ago.

The drift in nuclear DNA also has something to add to the story. Though humans are closely related to gorillas and chimpanzees, a comparison made between the drift in DNA between these different hominids reveals a striking difference. As separate populations, gorillas and chimpanzees are not geographically diverse but are each genetically diverse. Yet humans, who have spread across the globe, are genetically very closely related to each other. The only conclusion seems to be that all our ancestors died out, except for a single group from which all humans alive today are descended. This single group was no bigger than a few hundred people living in a single region of the world about 50,000 years ago, perhaps even earlier. It is thought that this group lived in East Africa, left together and migrated north-east, either up and across the Nile Delta or, more contentiously, by cutting directly across the Red Sea. Fifty thousand years ago the Red Sea was 70 metres shallower than it is today and narrower too. This small group thrived and their descendants populated the whole of the rest of the world over a period of a few tens of thousands of years, replacing and not interbreeding with other species in the genus *Homo* that they encountered. The trouble with this theory, though it is the most widely accepted of the modern theories, is that it is unable to account for the recent discovery that there are aboriginal populations alive today with distinct DNA that shows they do not have the common ancestors shared by the rest of the world.

Only in the last 40,000 years, in a process that took 15,000 years, did modern man spread as far as the region we now call Europe, coming from the Levant or, more controversially, from India.[10] Humans seem to have populated Australia long before they reached Europe. Evidence that supports the idea of early settlement in Australia is suggested by the sudden

---

[10] It has recently been argued that *Homo heidelbergensis* was replaced by *Homo sapiens* in India some 70,000 years ago. From there, *Homo sapiens* found their way to Australia and Europe between 50,000 to 40,000 years ago.

and mysterious disappearance of all animals in Australia weighing over 100 kilograms, which happened about 50,000 years ago. Placid large mammals would have been easy targets for hunters. There is no actual evidence that mankind was the cause of these extinctions, but there were humans around wherever these mass extinctions occurred across the globe. The Americas received their first human visitors just 11,000 years ago,[11] coinciding with the abrupt disappearance of 70 per cent of all large mammals in North America. After 33.7 million years of existence, the sabre-toothed tiger dies out about 9,000 years ago. After 4 million years of existence, the mastodon dies out 10,000 years ago. The Irish elk first appeared around 400,000 years ago and dies out 8,000 years ago. The mammoth, which was alive in the Pliocene epoch 4.8 million years ago, becomes extinct as recently as 4,500 years ago.

Why this small band of *Homo sapiens* that left Africa 50,000 years ago thrived, and how they replaced the other species in the *Homo* genus is not known. There is no evidence that violence was used, though our current nature undoubtedly makes this a strong hypothesis. One account, admitted to be merely a hunch on the part of its progenitor, the American psychologist Judith Rich Harris (b.1938), is that *Homo sapiens* hunted *Homo neanderthalensis* and ate them, taking their more hairy bodies as evidence of their animal status.[12] The trouble is we know less about the first *Homo sapiens* than we know about some other species in the genus *Homo*, certainly than we know about the Neanderthals. For now, a consensus of scientific opinion has *Homo neanderthalensis* as the last species in the *Homo* genus to overlap with the human species. A few years ago it was suggested that *Homo floresiensis* is more recent, and lived alongside *Homo sapiens* as recently as 12,000 years ago. There has been controversy surrounding access to the evidence which makes this attribution more contentious than

---

[11] A controversial theory argues that the Americas were settled as early as 33,000 years ago.
[12] It has also been suggested that *Homo sapiens* are cannibals.

is the norm even for this field. The counterargument is that the fossil evidence is not of a separate species but evidence of dwarfism or disease. The most recent population of Neanderthals we know of lived on the south coast of Gibraltar and died out about 30,000 years ago.

If increasing brain size is evidence of greater complexity evolving in the universe, then Neanderthals were the acme of creation so far, or at least from among the products of creation we know of, since it was they who had the largest brains of any species in the *Homo* genus known to date. Neanderthals were also thought to be stronger than we are.

But brain size and strength may just not have been enough. Nature doesn't select for brainiest but fittest. Nor does fittest mean strongest: it means best able to fit into the environment it inhabits. The British psychologist Nicholas Humphrey (b. 1943) has written about a species of monkey some of which are smart enough to find a way into a nut that is particularly difficult to crack. Unfortunately, the fruit inside turns out to be poisonous. In this case, the fittest monkeys in that population are the ones not quite smart enough to figure out how to get at the fruit.

Many significant cultural changes seem to have happened to the human population at or soon before the time of the last African exodus. Ritual burying and fish eating seem to be have been some of the first so-called cultural universals, something shared among all human tribes but not shared with any other species. Even when Neanderthals lived near rich sources of fish there is no evidence that they ever ate any. Ritual burial and fish eating seem to have emerged more than 110,000 years ago in the Levant. Art is another cultural universal. Excavations from 1991 at the Blombos cave in South Africa have unearthed engraved beads made from shells. Some 75,000 years old, these are the oldest examples of art[13] yet found. Cave paintings are harder to date. Although some cave paintings

---

[13] The bowerbird might object. Self-conscious art is what is meant, but then we have to ask what we mean by consciousness.

may be as old as 50,000 years, the oldest that can be definitively dated are 32,000 years old and are in France.

Early human tool use, which is stone-tool use, cannot be differentiated from Neanderthal tool use until about 50,000 years ago. Tool use had in fact remained unchanged for a couple of million years, but suddenly about 50,000 years ago tools became technologically more developed. Bone and antler-made tools also began to appear.[14] It is possible that language first began to develop in humans at this time, though naturally there are different theories, some of which argue for a gradual change over a much longer period of time (perhaps even millions of years). There is no agreement on whether Neanderthals possessed the ability to speak.

Over the next few tens of thousands of years modern humans begin to develop other cultural universals: religion, music, joke-telling, the incest taboo and cooking,[15] for example. The strategic conflict that we call war developed into the strategic conflict that we call games (or vice versa).

Civilisation seems to have begun in the Mediterranean, and in particular in the Levant. The Kebaran tribe of nomadic hunters and gatherers, the first anatomically modern humans, lived there from about 18,000 to 10,000 BC and were succeeded by the Natufian people.

Around 12,000 years ago the climate changed, and human culture changed forever. In a number of places, but perhaps first among these Natufian people, the warmer weather allowed farming to begin. Crops were cultivated, and animals domesticated. This was the first time that any species had begun to control the environment and the ecosystem. Mankind selected just a few plants and animals and began to change the world forever.

---

[14] There are various finds that upset this theory. Wooden spears discovered in the late 1990s in a coal mine at Schoeningen (100 kilometres east of Hanover in Germany) appear to date from some 400,000 years ago.

[15] Fire may have been around for 1 or 2 million years so cooking almost certainly precedes this time. Cuisine is what may have begun to emerge at this point.

We Are There

Sweet is the lore which Nature brings;
Our meddling intellect
Mis-shapes the beauteous forms of things:-
We murder to dissect.

Enough of science and of Art;
Close up these barren leaves;
Come forth, and bring with you a heart
That watches and receives.

William Wordsworth, 'The Tables Turned'

And the rest is history.

History comes to an end where it meets the present, the point at which the story meets the storyteller. *Now*, that axis about which swings the past and the future, is where we find ourselves. The laws of nature, however, describe a universe undifferentiated into a past and a future. As far as the fate of the universe is concerned, it is difficult to avoid the conclusion that if the human race disappeared tomorrow nothing would change. James Lovelock has predicted that billions of humans will have died by the end of the century as a direct result of global warming, and that Gaia may well find a new equilibrium for the planet that precludes human life. The American biologist Jared Diamond (b.1937) reminds us that in history all societies that have trashed their environments have become extinct, citing examples of civilisations that once flourished in Easter Island and in Greenland. In a few hundred million years' time the continents will have clumped together as they have done many times before. In a billion years' time the sun will be 10 per cent brighter than it is now. In 3 billion years' time the earth's iron core will have solidified. In 5 billion years' time the sun will run out of hydrogen and become a red giant. In billions of years' time our galaxy will have merged with neighbouring galaxy Andromeda. In tens of billions of years the billions of galaxies we currently see from the vantage point of earth (which by then will long since have been an unsuitable habitat for life) will have moved over the horizon of the visible universe. The night-time sky (whatever that means by then) will gradually empty out into complete blackness.

The long-term fate of the universe is described by theories at the furthest reaches of current scientific endeavour. These theories are necessarily speculative. What is perhaps surprising is how divergent the possible fates of the universe could be, and how sensitive they are to small changes in a small number of parameters. What is to become of the universe largely depends on how much mass and dark energy there is in it, which determines the universe's future rate of acceleration (or deceleration). Heat Death (also known as the Big Freeze) is thought to be the most likely outcome. The stars burn out. Galaxies collapse into black holes that then slowly evaporate. The universe becomes a soup of radiation that gradually cools towards absolute zero, the theoretical temperature at which atoms come as near to having no motion as the Heisenberg Uncertainty Principle allows. A universe that started out as radiation at the highest temperature possible, a quality of its deepest past, will end as radiation at the lowest temperature possible, a quality of its most distant future. Such a universe would be dominated by black holes 10^{40} years after the Big Bang, which might have all evaporated after another

1.7×10^{106} years. After that comes the Dark Era, which lasts forever.

If the many-worlds interpretation of quantum physics holds, then all possible fates of the visible universe are played out. And if the multiverse exists, out of which our local patch of visible universe emerged, then there are many other universes (perhaps an infinite number) to survive the death of this one, other universes in which the laws of nature may be quite different than they are here.

Although the universe as described by such stark physical laws appears to be oblivious to us, we may yet make our mark on it. The former Astronomer Royal, Martin Rees (b.1942), has predicted that mankind will find ways of tearing apart the fabric of space-time, of tearing the universe apart. Paradoxically, if we do manage to destroy the universe it could be argued that we will have proved to ourselves that we are uniquely privileged after all. If there were anti-Copernicans around to say it, they could say 'I told you so'. But even this violent act is not conclusive. The existence of the multiverse would mean that we had succeeded merely in destroying the local patch we call the visible universe. Our act would be reduced to utter insignificance. Once more, to contemplate the universe is to find ourselves at two poles at the same time: we are uniquely special and we are insignificant. The scientific method progresses by insisting on the insignificance, but repeatedly discovers privilege. To move forward scientists are forced to find ever more ingenious ways to re-establish our lack of consequence, and the universe answers in turn by refusing to disavow our centrality. It is unclear whether the game will ever come to an end. Yet we long for resolution. We want to believe that there is some ultimate answer to end our questioning. To believe that science will come to a full stop is to crave the certainty of an ending; it is to believe that there are laws of nature that fully describe the universe and that these laws can be found out.

But at a philosophical level there is something troubling about laws of nature that are written in stone. Why, when it undermines all other forms of uncertainty, should science believe that the laws of nature themselves are eternal? Galileo and Newton believed that God created the laws of nature and that it was the job of the scientist to uncover their workings. In this respect the scientific method flowed out of a belief system shared with monotheism: that there is something immutable and eternal about the universe. It was John Wheeler, some 30 years ago, who first began to question what it is that we mean by eternal laws and his ideas are back in fashion. He wondered whether in the future we might discover that the laws of nature themselves have evolved. What we call laws might have started off as something blurry and evolved through a process of natural selection into the laws as we find them to be today. And yet if the laws of nature evolve through natural selection, where does this put natural selection itself? Docs natural selection become the ultimate law of nature? Or is it something else, some sort of logical inevitability, an inescapable consequence, perhaps, of telling a story at all? For now, these are questions for philosophers. Pragmatic materialists will be prepared to wait for material answers.

Eternal laws do not contain any understanding of what it means to speak of the present, of the moment we call *now*. If the laws of nature are indeed eternal then our human experience of the present must be an illusion. A universe of eternal laws is laid out seamlessly. All that exists exists forever as an interconnected and inseparable web of phenomena. Einstein believed that his theories of relativity described such a reality: that the past and future are eternally existent and our egos fail to recognise it. A month before his own death, Einstein wrote about the recent death of his lifelong friend Michele Besso: 'Now he has departed from this strange world a little ahead of me. That signifies nothing. For us believing physicists the distinction between past, present, and future is only a stubbornly persistent illusion.' If this is the nature of reality, we might say that time does not flow, it just is. The appearance of time's arrow is an illusion of existence, an emergent quality that we experience at the scale of size we happen to inhabit in the universe, just as it is an illusion that time and space are separate if Einstein's space-time continuum exists.

If the present moment is not an illusion then it singles us out for privilege. Our experience of the here and now brings the story back to the storyteller. To dilute that privilege, and provide some comfort to Copernicans, we must hope that there are many other storytellers in other parts of the universe, spread across space and throughout time, telling the same story. But as Enrico Fermi once observed, if there are aliens, where are they? Should we be worried that none has been in touch, given that their existence is so often the materialists' defence against human privilege? One reason why we appear to have been unvisited is the vastness and extreme low density of space. A computer model made by the physicist Rasmus Björk in 2007 predicts that it would take 10 billion years to explore just 4 per cent of the universe, even if we could travel at a tenth the speed of light, a speed that is far beyond our abilities for the foreseeable future, and perhaps forever.

For the time being at least, the story of the material world remains our story alone. It begins with a description of a universe too simple to be fully apprehended, and ends with storytellers too complex to be fully described. In the balance, we place on the one side the universe as a whole, and in the other the human brain that conceives it. If it is something about our brain that appears to grant us privilege in the universe, we might wonder if there is any meaningful separation to be made between brain and universe. Ultimately we might wonder if there is any meaningful separation to be made between anything. We only appear special because we can't make a separation between the story and the storyteller. As Copernicans and materialists, however, the story continues because we ask ourselves: What is it to tell a story? And what might other storytellers look like? Is there something special about our brains that allows humans to find meaning in objects and symbols, and to organise the past and present into a conception of the future? If there is, then what is that something?

The human quest to understand the material world around

us presumably began on earth as soon as mankind became conscious. So what then do we mean by consciousness?

For more than 300 years the scientific method has divided the world into body and mind. Science took up the body and gave the mind and soul to religion. On one side is the world of subjective values, of aesthetics, morality and belief; on the other is science. 'Science got the better part of the bargain,' says the British biologist Rupert Sheldrake (b.1942), 'since it got practically everything.'

Although Descartes's philosophy of dualism has kept the machine-like body separate from a mind that cannot be described by physical laws, this rigid separation was not his intent. He meant to argue for the uniqueness of humans in the universe: that only humans possess this mysterious mind stuff. How mind could interact with body was a problem he failed to solve. He believed, incorrectly, that the pineal gland was a physical meeting point between these two worlds. Descartes's conception of the mind grants humans the kind of privilege that Copernicans must find distasteful, but by giving up mind for so long, science has been in danger of failing to account for it.

The American neurobiologist Roger W. Sperry (1913–1994) argued that consciousness is an emergent property, like the property of chairness that is inherent in the chair as we see it in the larger world, but disappears when we peer at the chair closely. Squinting, we see that the molecules from which the chair is constructed possess no qualities that we acknowledge in the larger world to be a chair. Sperry, who was the first to describe the separate functions of the left and right hemispheres of the brain, said that consciousness is like that: it is not reducible to physical processes, but emerges as a consequence of sufficient complexity in the workings of the brain, but disappears if that physical complexity is picked apart.

The scientific method has treated mind and matter as belonging to separate worlds, but the intimate connection between the body and the mind is clear to all living humans. As Nietzsche once wrote: 'The body is a great intelligence, a multiplicity with one sense, a war and a peace, a herd and a herdsman . . . There is more reason in your body than your best wisdom.' For now, it is in the brain rather than in the body as a whole that scientists have sought the mind. Recent direct investigations into mental states are beginning to show evidence of a physical connection between the brain and aspects of the world we once separated off as belonging to the domain of the mind. Electrical stimulation of the brain can change (for as long as the stimulation lasts) an individual's belief system; study of Buddhist monks shows that meditation changes physical structures in the brain, and in London taxi drivers the part of the brain related to memory is larger than it is for most of us.

What distinguishes a human brain from that of even our close relation the chimpanzee is the human brain's ability to make connections between cells. This malleability is peculiar to us. Here, then is evidence that mind has a material basis. Some scientists, however, have begun to wonder – admittedly it is a minority view – if the world is in fact the content of mind rather the other way about. 'I believe,' declares the cognitive philosopher Donald D. Hoffman, 'that consciousness and its contents are all that exists. Space-time, matter and fields never were the fundamental denizens of the universe but have always been among the humbler contents of consciousness dependent on it for their very being.' The essential character of material things, according to such a view, is that they are in some way a manifestation of consciousness, a view that Proust had already arrived at, and which he develops in his long novel *In Search of Lost Time* (1909–1922). 'Perhaps the immobility of the things around us is imposed on them by our certainty that they are themselves and not others, by the immobility of our mind confronting them.'

Even if we assume that the mind is an emergent property of the brain, only we humans among the life-forms we know of possess neural connections complex enough for the kind of consciousness to have emerged that allows us to comprehend the universe. We might even ask what the universe could mean if it were not observed by brains as complex as ours. 'It is impossible that there is a reality totally independent of the mind that conceives it, sees it, or senses it,' wrote the French mathematician Henri Poincaré (1854–1912). 'Even if it did exist, such a world would be utterly inaccessible to us.' If there was no comprehending consciousness in the universe, the great performance, as Schrödinger once said, plays to empty stalls. Is consciousness the universe becoming aware of itself for the first time? If it is, then, once more, it places a burden of privilege onto the brain and its emergent quality of mind (if that is indeed what we take the mind to be). The brain becomes paramount in a materialist description of the universe. And yet it is not easy to separate the brain even from the rest of the physical body that contains it, connected as the brain is to the nervous system. The body and its environment are seemingly inextricable. Perception appears to be a dynamic activity that cannot be broken down into parts. Rather than being separate from what we experience, the physical world itself becomes part of the experience. As Freeman Dyson puts it: 'Mind is woven into the fabric of the universe.' We are not apart from the world. The physical world is a manifestation of the act of perceiving it.

If the present moment that is you reading these words is real and is not predicted by the laws of nature, then we might take this as evidence that the scientific method is incomplete and will never fully describe the nature of reality. Scientists make measurements in the belief that if the measurement agrees with a prediction made by a theory then the measurement is of a world out there that exists independently of us. Out of the repeated affirmation of this external reality – the scientists' brand of faith – theory becomes enriched and measurement more refined. Science is a collection of facts and insights that follows the clue of its methodology, illuminates what we call the material world, and defines what we mean by progress. Whether or not science uncovers the truth doesn't come into it. Success is judged in its own terms, and is built on deep and unanswered mysteries (what is energy? what is a field?) that might be laid alongside the kinds of mysteries uncovered by the artist, philosopher and mystic. Materialists are tempted to look to the future where greater knowledge lies[1]; mystics, to the past where they see greater wisdom. But as Einstein warned: 'Whoever undertakes to set himself up as the judge in the field of truth and knowledge is shipwrecked by the laughter of the Gods.' Fortunately, there are scientists and artists happy to 'not dispute or assert but whisper results to his neighbour'.[2] Faith turns to dogma where there are sides, and where one side insists it sees Truth unseen by the other. And dogma leads to war: I am right, you are wrong, we are invincible, they are the enemy. Faith is 'the substance of things hoped for, the evidence of things not seen'.[3] Dogma is the insistence that what is unseen is really there. An alien viewpoint might have nothing more important to tell us than that what suits us might not suit others. 'Just try to get on,' we tell fighting children, from a perspective that might not be so different from an alien's.

Science and religion have appeared to be at war across the centuries. The history of science has been punctuated by a series of conflicts with the Church, but how those conflicts are interpreted is lopsided. We see the conflicts in hindsight, from the perspective of what the scientific method became. When Copernicus removed the earth from the centre of the cosmos, he began a process that needs to be understood as the historical beginnings of the scientific method, not merely as a reaction to the Church. Because it is built into its methodology, science is preoccupied with the idea of privilege. But science deflects attention away from its preoccupation by declaring that it is really a corrective to an ideology to which it is in opposition. It is true that the earth is at the centre of Aristotle's cosmos, but the Aristotelian earth is also at the bottom of the heap. It is the place to which earthly things –

---

[1] Though the American inventor Thomas Edison (1847–1931) is surely still correct in his assessment that 'we don't know a millionth of one per cent about anything'.
[2] In a letter written by the English poet John Keats.
[3] Hebrews 11:1.

things that contain the degraded stuff of the earth – fall. The earth is literally the place of fallen things. In the Church's adopted cosmology the earth was at the physical centre of the cosmos but this was not a privileged spot to be. The Church's opposition to the Copernican model was not made out of fear of downgrading the earth. It is the scientific method that was to equate physical centrality with privilege.

In that other great stand-off, Darwin certainly destroyed the notion that mankind stands at the head of a great chain of being; yet the notion that the earth and the animal kingdom exist in order to be exploited by a superior species is as implicit in the scientific method as it is in Judaeo-Christian doctrine. It might even be argued that science has extended the territory of exploitation to include the whole universe. Monotheistic religion and science both aim, one more explicitly than the other, to people the universe.[4] Science attempts and succeeds in making life more comfortable for some, but it also facilitates an increasing population and only partially provides the means to support it, and at ever greater cost to the planet. In time, science expects to people other planets across the universe. Indeed, it can have no other hope. Science and religion relieve suffering in the world but also increase suffering in the world. If religion often provides the reason for war, it is science that provides ever more sophisticated means of killing people.

When our desire is to conquer space, what can the nature of that desire be but to subjugate? Nature resists our attempts to uncover her secrets. Vast amounts of energy are required to reach into outer space, and vast amounts of energy are needed to breach the barrier of the atom. Unchecked, science and monotheism mean to vanquish Nature. But if we are to war against Nature we should not be surprised to find Nature enjoins the battle.

We cannot unravel the material world, and who would want

---

[4] The American astronomer Frank Drake (b.1930) once calculated that the energy of the sun could support 10^{22} human souls.

to? Materialism is the greatest story ever told. But we can try to understand what we mean by a material world and what we are in relationship to it. Materialism is the story of a universe that is full of meaning but without purpose. Whatever we mean by meaning is a local phenomenon that is centred on us. We, who are human observers living in the here and now, live in the midst of the universe; we appear to live midway between whatever we can see or mean by the largest and smallest things. It is hardly surprising, then, that meaning drains out of the universe at its edges. It must. It is we who define its edges. Meaning itself is inextricable from our own interaction with the world. Meaning that is independent of us is meaningless. The universe must look and mean something quite different seen from the perspective of, say, an atom. If we were aware of the smallest ticks of time, we might see nuclear processes evolve. At the other end of the spectrum of time, a different kind of consciousness might see plants grow, planets come and go, or galaxies collide and merge. If complexity seems to be most fully expressed in mid-sized bodies, it may only be because we think we are those mid-sized bodies. When we look out to the horizon, we are fooled into thinking that we are at the centre of all we survey.

Treating our universe of 10^{80} particles as a machine for processing information (measured in binary digits 0 or 1, known as 'bits') it has been estimated that 10^{120} bits of information has been processed so far, and that there are another 10^{120} bits of information left to process. As a processing machine, we might say the visible universe is half done. Otherwise, we live in the declining days of the universe's star making, in the early days of the universe's long decline. Scientists worry about what is to become of human life beyond even the long lifetime of the sun. But maybe *this* is our time in the universe. To be concerned about mankind's fate to the ends of time is a cover for that eternal dread: of our own mortality. We do not worry about the early days of the universe when we did not exist as humans, just as we do not worry about our non-existence before we were born. So why are we so exercised about what might become of us in the far reaches of time, except out of a vain desire to control the fate of the universe itself? We might be less bullying of the universe if we were more aware that the universe as we understand it is not separate from us. Try as we might, and however we describe it, we are inseparable from the universe. The universe is portable.

The possibility of an address in the universe depends on what we mean by the word *universe*, what we mean by *the* universe, and what we mean by *in* the universe. Where we are in the universe also depends on what we mean by *we*. As human beings *we* are separate egos looking out on a world of separate things. 'I am alone with the beating of my heart.' Science, however, is a collective experience of the world that makes the *we* ever more inclusive, and although science, too, starts off from the premise that the world is made of separate things, it progresses by unifying this separateness into a universe of inseparability. Science tells us that there is a *we* that is descended from a single mother who lived 150,000 years ago. Our DNA shows us that there is a *we* that is all living things all sharing the same DNA code: 3 billion years of evolving life. But why stop there? *We* are – everything is – woven out of primordial hydrogen that filled the universe around 14 billion years ago. Nor need we rest there. *We* are – everything is – evolved symmetrical radiation. And before that, *we* are something that is beyond whatever before can mean. I am here. You are there. We are everything and everywhere. They are us.

To desire an address is to desire status. But the notion of privilege becomes meaningless in a world that is all of a piece. Many scientists do not expect science ever to describe that wholeness, but the scientific methodology does at least point towards the unity of phenomena that we call the universe. Ultimately, this belief is not so very different from the direct apprehension of life we call mysticism, or which in other forms is the object of the life of the artist, philosopher or theologian. Scientists break up into pieces and lay reality along a line called progress. For the artist and the mystic, reality is all of a piece, time is circular, and the word progress is meaningless:

Picasso is not an advance on Titian (nor a diminution). But whatever way we come at reality, it's all a question of looking and of seeing. And even if science were one day to describe all that there is, unless it is to deny all other forms of truth-seeking, science and art and religion and philosophy must eventually meet up. As the American astronomer and physicist Robert Jastrow (1925–2008) has predicted: the scientist who has climbed the highest peak may find 'as he pulls himself over the final rock, [that] he is greeted by a band of theologians who have been sitting there for centuries'.

In a modern world obsessed with certainty and things eternal, we might learn to live in the uncertainty of an unending scientific process (without necessarily believing in unending scientific progress). We want to believe that things last forever, whether it is love, life, God, or the laws of nature. But death, as Freud continually reminds us, is what certainty looks like. Perhaps the best we can hope for is to live in uncertainty for as long as we can bear it.

Bibliography

Barnes, Jonathan, *Early Greek Philosophy* (Penguin Classics, 1987, revised 2001)

Barrow, John D., *The Constants of Nature* (Jonathan Cape, 2002)

Bohm, David, *Wholeness and the Implicate Order* (Routledge, 2002)

Brecht, Bertolt, *Life of Galileo*, trans. John Willets, Ralph Manheim, ed., (Penguin Classics, 2008)

Brockman, John, ed., *What We Believe but Cannot Prove* (Harper-Collins, 2006)

Bryson, Bill, *A Short History of Nearly Everything* (Doubleday, 2003)

Cadbury, Deborah, *The Dinosaur Hunters* (Fourth Estate, 2000)

Cadbury, Deborah, *The Space Race: The Battle to Rule the Heavens* (Fourth Estate, 2005)

Calaprice, Alice, *The Quotable Einstein* (Princeton University Press, 1996)

Charlesworth, Brian and Charlesworth, Deborah, *Evolution* (Oxford University Press, 2003)

Cheetham, Nicholas, *Universe: A Journey from Earth to the Edge of the Cosmos* (Smith Davies, 2005)

Coles, Peter, *Cosmology* (Oxford University Press, 2001)

Conze, Edward, *Buddhist Wisdom Books* (Allen and Unwin, 1958)

Cook, Michael, *A Brief History of the Human Race* (Granta Books, 2004)

Dalai Lama, His Holiness the, *The Universe in a Single Atom* (Morgan Road Books, 2005)

Dart, Raymond, *Adventures With the Missing Link* (Hamish Hamilton, 1959)

Darwin, Charles, *On the Origin of Species* (John Murray, 1859)

Darwin, Charles, *The Descent of Man* (John Murray, 1871)

Davies, Merryl Wyn, *Darwin and Fundamentalism* (Icon Books, 2000)

Davies, Paul, *Other Worlds* (J M Dent, 1980)

Davies, Paul, *God and the New Physics* (J M Dent, 1983)

Dawkins, Richard, *The Selfish Gene* (Oxford University Press, 1976)

Dawkins, Richard, *The Blind Watchmaker* (Longman, 1986)

Delsemme, Armand, *Our Cosmic Origins* (Cambridge University Press, 1998)

Dennett, Daniel C., *Darwin's Dangerous Idea* (Simon and Schuster, 1995)

Diamond, Jared, *Collapse: How Societies Choose to Fail or Survive* (Viking Books, 2005)

Dressler, Alan, *Voyage to the Great Attractor: Exploring Intergalactic Space* (Knopf, 1994)

Eccles, John C., *The Human Mystery* (Springer, 1979)

Eldredge, Niles, *The Triumph of Evolution and the Failure of Creationism* (Henry Holt, 2000)

Eliot, T. S., *Collected Poems 1909–1962* (Faber and Faber Ltd, 1963); lines from 'The Love Song of J Alfred Prufrock' and 'The Waste Land' quoted by permission of the estate of T. S. Eliot and Faber and Faber.

Fergusson, Kitty, *The Fire in the Equations: Science, Religion and the Search for God* (Bantam Press, 1994)

Ferreira, Pedro G., *The State of the Universe* (Weidenfeld and Nicolson, 2006)

Feynman, Richard P., *Surely You're Joking Mr Feynman* (W W Norton, 1985)

Feynman, Richard P., *The Meaning of It All* (Addison-Wesley, 1998)

Fisher, Len, *Weighing the Soul* (Weidenfeld and Nicolson, 2004)

Forbes, Peter, *The Gecko's Foot: Bio-inspiration, Engineering New Materials from Nature* (Fourth Estate, 2006)

Fortey, Richard, *The Earth* (HarperCollins, 2004)

Gee, Henry, *Deep Time* (Fourth Estate, 2000)

Gee, Henry, *Jacob's Ladder* (Fourth Estate, 2004)

Gleick, James, *Chaos* (William Heinemann Ltd, 1988)

Gleick, James, *Genius: Richard Feynman and Modern Physics* (Little, Brown, 1992)

Gleick, James, *Isaac Newton* (Fourth Estate, 2003)

Gould, Stephen Jay, *Wonderful Life* (W W Norton, 1989)

Gould, Stephen Jay, *Bully for Brontosaurus* (W W Norton, 1991)

Grant, Edward, *Physical Science in the Middle Ages* (John Wiley and Sons, 1971)

Gray, John, *Straw Dogs* (Granta Books, 2002)

Greene, Brian, *The Elegant Universe* (Vintage, 2000)

Gribbin, John, *Science: A History* (Allen Lane, 2002)

Guth, Alan, H., *The Inflationary Universe* (Jonathan Cape, 1997)

Haldane, John, *An Intelligent Person's Guide to Religion* (Duckworth Overlook, 2003)

Hall, A. Rupert, and Hall, Marie Boas, *A Brief History of Science* (The New American Library, 1964)

Hawking, Stephen, *A Brief History of Time* (Bantam Press, 1988)

Haxton, Brooks, trans., *Heraclitus Fragments* (Viking, 2001)

Haynes, Jane, *Who Is It That Can Tell Me Who I Am?* (foreword by Hilary Mantel, intheconsultingroom.com, 2006)

Hoffman, Paul, *The Man Who Loved Only Numbers* (Hyperion, 1998)

Hughes, Ted, *Tales from Ovid* (Faber and Faber Ltd, 1997). Lines quoted by permission of the estate of Ted Hughes and Faber and Faber.

Jastrow, Robert, *God and the Astronomers* (W W Norton, 1978)

Jung, C. G., *Memories, Dreams, Reflections*, recorded and edited by A. Jaffe (Collins, 1962)

Jung, C. G., *Synchronicity* (1952; Princeton University Press, 1973)

Kirk, G. S., and Raven, J. E., *The Presocratic Philosophers* (Cambridge Univeristy Press, 1957)

Kuhn, Thomas S., *The Structure of Scientific Revolutions* (University of Chicago Press, 1962)

Mann, Thomas, *The Magic Mountain*, trans. John E. Woods (Everyman's Library, 2005).

May, Brian; Moore, Patrick and Lintott, Chris, *Bang: The Complete History of the Universe* (Carlton, 2006)

Monod, Jacques, *Chance and Necessity: Essay on the Natural Philosophy of Modern Biology*, trans. A. Wainhouse (Collins, 1972)

Moring, Gary F., *The Complete Idiot's Guide to Theories of the Universe* (Alpha Books, 2002)

Nemiroff, Robert J., and Bonnell, Jerry T., *The Universe: 365 Days* (Harry N. Abrams, 2003)

Newton, Roger G., *Galileo's Pendulum* (Harvard University Press, 2004)

Nietzsche, Friedrich, *The Birth of Tragedy*, trans. Shaun Whiteside (Penguin Classics, 1993)

Nietzsche, Friedrich, *Twilight of the Idols*, trans. R. J. Hollingdale (Penguin Classics, 1990)

Oerter, Robert, *The Theory of Almost Everything* (Pi Press, 2006)

Panek, Richard, *Seeing and Believing* (Viking, 1998)

Panek, Richard, *The Invisible Century: Einstein, Freud and the Search for Hidden Universes* (Viking, 2004)

Penrose, Roger, *The Road to Reality* (Knopf, 2004)

Phillips, Adam, *Darwin's Worms* (Faber and Faber Ltd, 1999)

Polkinghorne, J. C., *The Quantum World* (Longman, 1984)

Popper, Karl, *The Logic of Scientific Discovery* (English translation: Hutchinson, 1959)

Popper, Karl R., and Eccles, John C., *The Self and Its Brain* (Springer, 1977)

Primack, Joel R., and Abrams, Nancy Ellen, *The View from the Center* (Riverhead Books, 2006)

Proust, Marcel, *In Search of Lost Time* (Allen Lane, 2002)

Ramsey, Frank, *The Foundations of Mathematics and Other Logical Essays* (London, 1931)

Randall, Lisa, *Warped Passages: Unravelling the Mysteries of the Universe's Hidden Dimensions* (HarperCollins, 2005)

Reader, John, *Missing Links* (Little, Brown, 1981)

Redfern, Martin, *The Earth: A Very Short Introduction* (Oxford University Press, 2003)

Rees, Martin, *Just Six Numbers* (Basic Books, 2001)

Rees, Martin, *Our Final Hour* (Basic Books, 2003)

Ricard, Matthieu, and Thuan, Trinh Xuan, *The Quantum and the Lotus* (Three Rivers Press, 2001)

Ridley, Matt, *Genome* (Fourth Estate, 1999)

Ridley, Matt, *Nature via Nurture* (Fourth Estate, 2003)

Rollins, Hyder E., ed., *The Letters of John Keats* (Harvard University Press, 1958)

Seife, Charles, *Decoding the Universe* (Viking, 2006)

Sheldrake, Rupert, *A New Science of Life* (J P Tarcher, 1982)

Singh, Simon, *Big Bang* (Fourth Estate, 2004)

Sobel, Dava, *Galileo's Daughter* (Fourth Estate, 1999)

Sobel, Dava, *The Planets* (Fourth Estate, 2005)

Swain, Harriet, ed., *Big Questions in Science* (introduction by John Maddox, Jonathan Cape, 2002)

Taylor, Timothy, *The Prehistory of Sex: Four Million Years of Human Sexual Culture* (Fourth Estate, 1996)

Taylor, Timothy, *The Buried Soul: How Humans Invented Death* (Fourth Estate, 2002)

Weinberg, Steven, *The First Three Minutes* (Andre Deutsch, 1977, revised edition Basic Books, 1993)

Wilbur, Richard, *New and Collected Poems* (Faber and Faber Ltd, 1989). Lines from 'Epistemology' quoted by permission of Richard Wilbur and Faber and Faber.

During the writing of this book I frequently turned to the online resource Wikipedia. It has its detractors, but I am not one of them. I found it to be consistently accurate, and invariably up to date. I made use of websites too numerous to list in full, but exceptional among them are the particle tour devised by the Particle Data Group of Lawrence Berkeley National Laboratory (http://particleadventure.org/) and NASA's website (www.nasa.gov). I also owe a huge debt of gratitude to the weekly journal *New Scientist.*

Acknowledgments

This book would have been less error-free were it not for the generosity, advice, and expertise of Andrew Coleman, Stacey D'Erasmo, Peter Forbes, Meg Giles, Tim Hughes, Kate Jennings, Daniel Kaiser, Gerald McEwen, Hilary Mantel, Graeme Mitchison, Cynthia O'Neal, Richard Panek, Seth Pybas, Matt Ridley, Steven Rose, Simon Singh, Dava Sobel, and Timothy Taylor, who each read this book at some stage in its gestation.

The writing of this book would have been a less pleasurable experience, and might never have got going, had it not been for the encouragement of Gillon Aitken, Jason Arbuckle, Thomas Blaikie, Carol Bosiger, Melanie Braverman, Bill Clegg, Hazel Coleman, Michael Cunningham, Michael Gormley, Courtney Hodell, Sarah Lutyens, Blue Marsden, Kathleen Ollerenshaw, Shabir Pandor, Molly Perdue, Beth Povinelli, Noni Pratt, Sally Randolph, Joyce Ravid, Jana Warchalowski, and Cathy Westwood.

I am forever in the debt of my agent Michael Carlisle.

The publication process was much enhanced by the contributions of Stephen Appleby, Ethan Bassoff, Tess Callaway, Tim Duggan, Sue Freestone, Caroline Gascoigne, Eddie Mizzi, James Nightingale, Susan Sandon, and Michael Schellenberg.

This book would not exist were it not for the love and support of Jane Haynes, James Lecesne, Peter Parker, and Salley Vickers.

And most apparently, this book would not exist were it not for my mother, to whom this book is dedicated.

Index